Daniela Diessel

Near-field Optical Spectroscopy on Photonic Metamaterials

Daniela Diessel

Near-field Optical Spectroscopy on Photonic Metamaterials

Dissertation at the Karlsruhe Institute of Technology

Südwestdeutscher Verlag für Hochschulschriften

Imprint
Any brand names and product names mentioned in this book are subject to trademark, brand or patent protection and are trademarks or registered trademarks of their respective holders. The use of brand names, product names, common names, trade names, product descriptions etc. even without a particular marking in this work is in no way to be construed to mean that such names may be regarded as unrestricted in respect of trademark and brand protection legislation and could thus be used by anyone.

Publisher:
Südwestdeutscher Verlag für Hochschulschriften
is a trademark of
Dodo Books Indian Ocean Ltd., member of the OmniScriptum S.R.L Publishing group
str. A.Russo 15, of. 61, Chisinau-2068, Republic of Moldova Europe
Printed at: see last page
ISBN: 978-3-8381-1240-4

Zugl. / Approved by: Karlsruhe, KIT, Diss., 2011

Copyright © Daniela Diessel
Copyright © 2011 Dodo Books Indian Ocean Ltd., member of the OmniScriptum S.R.L Publishing group

Contents

1	**Introduction**	**1**
2	**Fundamentals**	**5**
2.1	Electrodynamics	5
	2.1.1 Maxwell's equations	5
	2.1.2 Material parameters	6
	2.1.3 Electromagnetic waves	7
	2.1.4 Drude model	8
2.2	Metamaterials	9
	2.2.1 Split-ring resonators	10
	2.2.2 Double-wires	13
	2.2.3 Fishnet structure	15
	2.2.4 Magnetic interactions between artificial atoms	17
2.3	Calculation tools	18
	2.3.1 Finite-integration technique	19
	2.3.2 Scattering-Matrix Method	20
2.4	Near-field microscopy	21
	2.4.1 Resolution limit	21
	2.4.2 Resolution enhancement in the near-field	22
	2.4.3 Experimental implementation of near-field measurements	25
3	**Experimental methods**	**29**
3.1	Probe fabrication	29
	3.1.1 Chemical etching process	29
	3.1.2 Metallisation and aperture cutting	31
	3.1.3 Probe module	32
3.2	The near-field microscope	34
	3.2.1 Optical setup	35
	3.2.2 SNOM unit	36
3.3	Sample fabrication	37
3.4	Transmittance characterisation	40

3.5 Phase-sensitive near-field microscope 41

4 Model for the near-field probe 43
4.1 Theoretical description of the imaging process 43
4.2 Numerical implementation 48
    4.2.1 Field distribution calculations 48
    4.2.2 Transmission through the probe 49
4.3 Exemplary results 51

5 Double-wire structure 55
5.1 Preparatory examinations 55
5.2 Frequency-dependent investigations 57
    5.2.1 Near-field measurements 58
    5.2.2 Calculations 59
    5.2.3 Spectral behaviour 60

6 Fishnet structure 63
6.1 Far-field properties 63
6.2 Near-field investigations 64
    6.2.1 Near-field distribution 64
    6.2.2 Spectral behaviour 66
    6.2.3 Transition to the far-field 67

7 Low-symmetry split-ring-resonator arrays 69
7.1 Far-field properties 69
7.2 Near-field investigations 72
    7.2.1 Diagonal polarisation 72
    7.2.2 Horizontal and vertical polarisation 74

8 Conclusion 77

Bibliography 81

Publications 89

Acknowledgements 91

1 Introduction

Photonic metamaterials are a recently devised type of artificial material composed of periodically arranged, nanoscopic metallic building blocks. These building blocks, the "meta-atoms", interact with electromagnetic waves and with each other, rendering possible novel optical properties such as, for instance, magnetic responses at optical frequencies, negative permittivities and permeabilities, and even negative refractive indices. With these characteristics metamaterials expanded the range of what can be realised in optics, stimulating a vibrant and fast-growing field of fundamental research.

The initial impulse for this development came from Victor Veselago who demonstrated in a theoretical study in 1964, that the refractive index of a material can be negative if both the permittivity (ε_r) and permeability (μ_r) of that material take on values below zero [1]. While negative ε_r occur naturally, media with permeabilities different from one at optical frequencies were not known at that time. This changed in 1999, when Pendry et al. proposed a structure design based on the concept of the LC circuit, namely, periodically arranged coils, the ends of which act as capacitors [2]. When illuminated by an electromagnetic wave (within the appropriate frequency range), an oscillating current is resonantly induced in the coils, and opposite charges alternately accumulate in the capacitors. This creates the (negative) magnetic response of the structure. A simplified version of this design, the split-ring resonator, is one of today's most commonly used "meta-atoms", that can for example be used to study magnetic interactions at optical frequencies between the individual "atoms" [3]. A negative magnetic response can also be obtained with the double-wire structure, the "atoms" of which consist of two stacked strips of metal separated by a spacer layer [4]. If a negative electric response in the form of a diluted metal—i.e., metal strips at right angles to the double-wires—is added to this design, the combined structure exhibits both a negative ε_r and μ_r and, thus, a negative refractive index [5]. Due to its grid-like appearance, this metamaterial is called "fishnet structure". Effects achievable with negative refractive indices include negative refraction [6], superlenses [7] and reversed Cherenkov radiation [8]. Another exciting new possibility offered by metamaterials is electromagnetic cloaking through use of transformation optics [9, 10].

The metamaterials' unusual properties are generated by a complex interplay of electromagnetic fields and currents. In order to further our understanding of the properties, this interplay has to be investigated more closely. Thus, it would be desirable to experimentally observe the distribution of the electromagnetic fields. However, as the dimensions of the "meta-atoms" and their separations are by definition far smaller than the wavelength at which they function, the resolution limit of classical optics ($\approx \lambda/2$) prevents a direct determination of the fields with a conventional microscope. This is due to the fact that high spatial detail is encoded in electromagnetic modes with high wavenumbers. As only modes with wavenumbers below that of free space can radiate, the amount of detail that can be picked up at a distance is limited. This can be circumvented by using a scanning near-field microscope (SNOM) which picks up the non-radiating *near-fields*. The principle of this measurement technique was introduced by Edward Synge in 1928 [11]: a nanoscopically small aperture in an otherwise opaque, planar film is scanned closely above the examined structure. In this way the near-fields at the location of the aperture are scattered and transmitted by the aperture, so that they can be picked up with a detector placed behind the opaque film. The resolution achieved with this method depends on the size of the aperture and its distance to the sample. In 1944 and 1950, respectively, Hans Bethe and Christoffel Bouwkamp analytically described the transmission through a small hole in a planar screen, [12] and [13]. The first experimental realisation at a wavelength of 3 cm was published in 1972 by Ash and Nicholls [14], the first at optical wavelengths in 1984 by Pohl *et al.* [15]. Today, most aperture probes consist of an aperture at the tip of metallised optical fibres, introduced by Betzig *et al.* [16], which are tapered by means of heating and pulling [16, 17] or by a chemical etching process [18]. The distance to the sample is controlled by a feed-back method with piezoelectric shear-force detection, established by Karrai and Grober [19]. An alternative way of measuring the near-fields utilises nanoscopically sharp, apertureless tips that scatter the near-fields towards a detector (apertureless SNOM) [20].

Photonic metamaterials were first studied with such an apertureless near-field microscope by Zentgraf *et al.* in 2008 [21]. They investigated split-ring resonators and found that the signal obtained is mainly composed of the electric field component perpendicular to the sample surface. To also detect the in-plane components, aperture-SNOM has to be used. In the THz regime near-field distributions have been measured by Bitzer *et al.* [22]. For comparison with theory, however, only the calculated current densities were shown there. While their calculation is straightforward, the numerical derivation of the measured signal is highly involved. Approaches to theoretically reproduce the imaging process generally employ simplifications such as a restriction to two dimensions [23, 24]. Another possibility is to disregard interactions between probe and sample [25]. Then a transfer function can be derived, as performed by Bozhevolnyi *et al.* for a near-field microscope working with total internal reflection illumination [26].

In the course of this work, this approach was used for a collection-mode SNOM. A scheme was created to model the near-field microscope's imaging process, which allows an easy and fast numerical derivation of SNOM signals. Images obtained with this model were compared to experimental data of specifically designed and fabricated samples. We conducted measurements of the near-field signal intensity, finding effects of magnetic interactions between "meta-atoms", and investigated phase and amplitude distributions as a function of wavelength and as a function of the distance between probe and sample. Thus, the near-field distribution above several metamaterial designs (the double-wire structure, the fishnet structure, and the low-symmetry split-ring-resonator array) was measured with an aperture SNOM and compared to theoretically derived data calculated with a specifically constructed model.

Outline of this thesis

In chapter 2 we present the theoretical prerequisites for this thesis: we recapitulate the fundamentals of electrodynamics, discuss both the concept of metamaterials and the characteristics of the metamaterials studied in this work, namely the double-wire structure, the fishnet and the low-symmetry split-ring-resonator array. Then, we outline the algorithms used for the numerical calculation of the near-fields in these samples and introduce the principles and techniques of scanning near-field optical microscopy. Following this, the experimental methods employed in this work are detailed in chapter 3. We begin the chapter with the fabrication of near-field probes and the setup of our near-field optical microscope. Subsequently, the processes involved in manufacturing the metamaterial samples and determining their far-field properties are summarised. We end by describing the phase- and polarisation-sensitive, heterodyne near-field microscope employed for near-field examinations of the double-wire and fishnet structures.

Our model for the imaging process of near-field aperture probes is explicated in chapter 4 and substantiated by a numerical study. After this, its implementation is detailed and results are exemplarily compared to near-field measurements conducted earlier in our group.

We investigate the counter-intuitive distribution of the near-field signal of double-wire structures in chapter 5, by examining samples with different wire spacing. Then, frequency-dependent measurements are presented and are taken as a basis to explore the limitations of our numerical model with respect to variations of wavelength. We take advantage of the double-wires' symmetry characteristics to obtain the theoretical, frequency-dependent signal. In chapter 6, experimental data for the fishnet structure are shown and compared to calculated data. The near-field distribution and its change with wavelength is examined, as well as the transition from the near-field to the far-field domain. Finally, in chapter 7, we visualise and analyse the effects of magnetic interactions between the individual resonators of low-symmetry split-ring-resonator arrays, again supporting our conclusions with calculated images.

We conclude the thesis in chapter 8 with a summary of our results and an outlook.

2 Fundamentals

In the following chapter the theoretical prerequisites for this work are recapitulated. We begin by summarising some of the fundamentals of electrodynamics (chapter 2.1). Then, the concept of metamaterials will be introduced, as well as the numerical approaches used in this thesis to calculate the electromagnetic properties of these materials (chapters 2.2 and 2.3, respectively). Finally, we give a brief overview over the basic theoretical principle of near-field microscopy and the most commonly employed methods in this field (chapter 2.4).

2.1 Electrodynamics

An overview over the theoretical fundamentals of electromagnetic interactions and optics is given in the following sections. The equations reviewed here are mostly general knowledge and can be found in a great number of books. We recommend [27] for basic reading and [28] for a more specific treatment of the matter of nano-optics.

2.1.1 Maxwell's equations

The interaction of the electric and magnetic fields with one another and with the charged matter around them are fully described by the well-known set of the *macroscopic Maxwell's equations*, given below. Depending on the requirements of the studied systems, they are expressed either in their differential (left-hand side) or in their integral form (right-hand side):

$$\nabla \cdot \boldsymbol{D} = \varrho \qquad \oint_A \boldsymbol{D} \, \mathrm{d}\boldsymbol{A} = \int_V \varrho \, \mathrm{d}V, \qquad (2.1\mathrm{a})$$

$$\nabla \cdot \boldsymbol{B} = 0 \qquad \oint_A \boldsymbol{B} \, \mathrm{d}\boldsymbol{A} = 0, \qquad (2.1\mathrm{b})$$

$$\nabla \times \boldsymbol{E} = -\dot{\boldsymbol{B}} \qquad \oint_{\delta A} \boldsymbol{E} \, \mathrm{d}\boldsymbol{s} = -\int_A \dot{\boldsymbol{B}} \, \mathrm{d}\boldsymbol{A}, \qquad (2.1\mathrm{c})$$

$$\nabla \times \boldsymbol{H} = \boldsymbol{j} + \dot{\boldsymbol{D}} \qquad \oint_{\delta A} \boldsymbol{H} \, \mathrm{d}\boldsymbol{s} = \int_A \boldsymbol{j} \, \mathrm{d}\boldsymbol{A} + \int_A \dot{\boldsymbol{D}} \, \mathrm{d}\boldsymbol{A}, \qquad (2.1\mathrm{d})$$

As usual, $\boldsymbol{E}$ and $\boldsymbol{H}$ are, respectively, the electric and magnetic fields, $\boldsymbol{D}$ is the electric displacement field, and $\boldsymbol{B}$ the magnetic induction field. ϱ denotes the external charge density and $\boldsymbol{j}$ the external current density.

In this work both versions of Maxwell's equations are used. In the following sections, the differential form is employed to discuss the basic electromagnetic properties of matter and to derive the wave equations with the plane-wave solution. The integral form is incorporated in the numerical finite-integration solver described in chapter 2.3.1, which is utilised for the calculations in this thesis.

2.1.2 Material parameters

Note that the internal charge and current densities appear only implicitly in Eqs. (2.1). In most materials the charges and currents react uniformly to externally applied fields. They can thus be considered as a whole. Respectively, they give rise to the electric susceptibility χ_e and the magnetic susceptibility χ_m of the material. Through the polarisation $\boldsymbol{P}$ and magnetisation $\boldsymbol{M}$ they form part of the electric displacement $\boldsymbol{D}$ and the magnetic induction $\boldsymbol{B}$ (all depending on the frequency of the applied fields):

$$\boldsymbol{D} = \varepsilon_0 \boldsymbol{E} + \boldsymbol{P} = \varepsilon_0 \boldsymbol{E} + \varepsilon_0 \chi_e \boldsymbol{E}, \tag{2.2a}$$

$$\boldsymbol{B} = \mu_0 \boldsymbol{H} + \mu_0 \boldsymbol{M} = \mu_0 \boldsymbol{H} + \mu_0 \chi_m \boldsymbol{H}. \tag{2.2b}$$

Here, as usual, μ_0 is the vacuum permeability defined as $4\pi \cdot 10^{-7}\,\text{Hm}^{-1}$, and $\varepsilon_0 = 1/\mu_0 c_0^2$ is the vacuum permittivity, with the free-space velocity of light defined as $c_0 = 299\,792\,458\,\text{ms}^{-1}$. With the relative permittivity $\varepsilon_r = (1 + \chi_e)$ and the relative permeability $\mu_r = (1 + \chi_m)$ Eqs. (2.2) can be reduced to give:

$$\boldsymbol{D} = \varepsilon_0 \varepsilon_r \boldsymbol{E}, \tag{2.3a}$$

$$\boldsymbol{B} = \mu_0 \mu_r \boldsymbol{H}. \tag{2.3b}$$

This approach tacitly averages over the microscopic distribution and movement of the "bound" charges and the spaces between them, treating the medium as a *homogeneous material*. Obviously, this can only be done because the exact distribution of the fields below the corresponding length scale is of no import. As will be discussed in chapter 2.2, an effective relative permittivity and permeability can be assigned even to whole assemblies of different homogeneous materials. There, too, the assigned effective material parameters only hold meaning on length scales well above the averaging length.

2.1 Electrodynamics

Importantly, the reaction of the internal charges depends on the frequency of the externally applied fields. Thus, $\varepsilon_r = \varepsilon_r(\omega)$ and $\mu_r = \mu_r(\omega)$. It should be mentioned that the shown approach applies to *anisotropic* materials as well. Here, the relative permittivity and permeability can depend on the direction, so that ε_r, μ_r, χ_e and χ_m become second-rank tensors. Moreover, in *bi-isotropic* and *bi-anisotropic* materials the electric field can induce a magnetic response and the magnetic field an electric response. In such materials, a coupling parameter ξ has to be introduced, and Eqs. (2.2a) and (2.2b) have to be extended by $+\xi H$ and $-\xi^T E$, respectively [29]. As a consequence, Eqs. (2.3) then become:

$$D = \varepsilon_0 \varepsilon_r E + \xi H, \quad (2.4a)$$

$$B = \mu_0 \mu_r H - \xi^T E. \quad (2.4b)$$

Again, in the anisotropic case ε_r, μ_r and ξ are tensors.

Finally, most materials do not exhibit any magnetic response at optical frequencies, for which reason their magnetic susceptibility vanishes and $\mu_r = 1$. In metamaterials, however, the *effective* permeability can well assume different values (see chapter 2.2).

2.1.3 Electromagnetic waves

For media without external charges and currents, ϱ and j vanish. If, for such media, we take the curl of Eqs. (2.1c) and (2.1d) and use the identity $\nabla \times \nabla \times = -\nabla^2 + \nabla\nabla\cdot$, we obtain:

$$-\nabla^2 E + \varepsilon_r \mu_r \frac{1}{c_0^2} \ddot{E} = 0, \quad (2.5a)$$

$$-\nabla^2 H + \varepsilon_r \mu_r \frac{1}{c_0^2} \ddot{H} = 0. \quad (2.5b)$$

These are the *wave equations* for homogeneous media without free charges and currents. They can be solved with the ansatz for a monochromatic *plane wave* with the *wave vector* k and the *angular frequency* ω:

$$E(r,t) = E_0 e^{i(kr - \omega t)}, \text{ and} \quad (2.6a)$$

$$H(r,t) = H_0 e^{i(kr - \omega t)}. \quad (2.6b)$$

When inserted into the wave equations, Eqs. (2.5), this leads to the *dispersion relation*

$$|k| = |\pm \sqrt{\varepsilon_r \mu_r}| \cdot \frac{\omega}{c_0}, \quad (2.7)$$

and the *refractive index* n

$$n = \pm\sqrt{\varepsilon_r \mu_r}. \tag{2.8}$$

As noted by Victor Veselago, this refractive index can have negative values, especially if both ε_r and μ_r are negative [1]. A material with $n < 0$ is called a *negative-index material*. For all passive media, the sign of the index, *i.e.*, the sign of the root $\sqrt{\varepsilon_r \mu_r}$, is dictated by the dissipation of energy: the imaginary part of n cannot be negative, as this would lead to a growing amplitude of the plane wave in the Eqs. (2.6). Squaring Eq. (2.8), and using $n = n_1 + in_2$, $\varepsilon_r = \varepsilon_1 + i\varepsilon_2$, and $\mu_r = \mu_1 + i\mu_2$ leads to

$$n^2 = (n_1 + in_2)^2 = n_1^2 - n_2^2 + i(2n_1 n_2) \tag{2.9a}$$
$$= (\varepsilon_1 + i\varepsilon_2) \cdot (\mu_1 + i\mu_2) = \varepsilon_1 \mu_1 - \varepsilon_2 \mu_2 + i(\varepsilon_1 \mu_2 + \varepsilon_2 \mu_1). \tag{2.9b}$$

As the imaginary part of n has to be positive, that is to say $n_2 > 0$, the material parameters have to meet the following condition for the real part of the refractive index to be negative:

$$\mathrm{Re}(n) < 0 \quad \Leftrightarrow \quad \varepsilon_1 \mu_2 + \varepsilon_2 \mu_1 < 0 \tag{2.10}$$

With this condition the sign of n is determined unequivocally.

Finally, as a wave propagates through a material, it acquires a phase $\varphi = \boldsymbol{k}\boldsymbol{r}$. The sign of this phase depends on that of the wave vector k. Hence, at a given time, φ increases with distance in a medium with $n > 0$, whereas it decreases if $n < 0$. Thus, propagation through a negative-index material leads to a negative phase.

2.1.4 Drude model

In the year 1900 Paul Drude proposed a model to derive the conductivity and other properties of metals [30]. With this approach the permittivity can be deduced quite accurately. Following his model, the metal is described as a lattice of stationary ions with electrons moving freely between them. The electrons have an effective mass m_e^* determined by the band structure. As they frequently collide with the ions, a damping factor ω_c—the collision frequency—is introduced. They are driven by an external electric field $\boldsymbol{E} = \boldsymbol{E}_0 \cdot e^{-i\omega t}$. The equation of motion thus reads:

$$m_e^* \ddot{\boldsymbol{x}} + m_e^* \omega_c \dot{\boldsymbol{x}} = -e\boldsymbol{E}_0 \cdot e^{-i\omega t}. \tag{2.11}$$

An ansatz $\boldsymbol{x} = \boldsymbol{x}_0 e^{-i\omega t}$ leads to a polarisation

$$\boldsymbol{p} = -e\boldsymbol{x} = \frac{-e^2}{m_e^* \omega(\omega + i\omega_c)} \boldsymbol{E}_0 e^{-i\omega t} \equiv \alpha \boldsymbol{E}, \tag{2.12}$$

2.2 Metamaterials

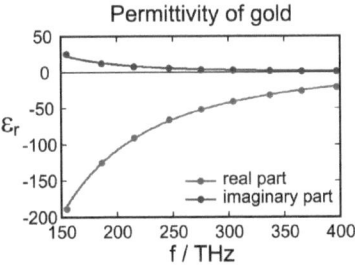

Fig. 2.1: Real part (red) and imaginary part (blue) of the permittivity of gold. The continuous lines represent theoretical data calculated with the Drude model ($\omega_{\mathrm{pl}} = 1.37 \cdot 10^{16}\,\mathrm{s}^{-1}$ and $\omega_{\mathrm{c}} = 1.12 \cdot 10^{14}\,\mathrm{s}^{-1}$), the circles indicate experimental data from [31].

where α is the polarisability. With the electron density n_0 and the electric polarisation $\boldsymbol{P} = n_0 \boldsymbol{p} = \varepsilon_0 \chi \boldsymbol{E}$ it follows that $\chi = n_0 \alpha / \varepsilon_0$, and thus we obtain the permittivity,

$$\varepsilon_{\mathrm{r}} = 1 + \chi = 1 - \frac{\omega_{\mathrm{pl}}^2}{\omega(\omega + i\omega_{\mathrm{c}})}, \qquad (2.13)$$

where the plasma frequency is introduced:

$$\omega_{\mathrm{pl}} = \sqrt{\frac{n_0 e^2}{\varepsilon_0 m_{\mathrm{e}}^*}}. \qquad (2.14)$$

The real and imaginary parts of a permittivity derived with this model are shown in Fig. 2.1 (continuous lines). Here, a plasma frequency of $\omega_{\mathrm{pl}} = 1.37 \cdot 10^{16}\,\mathrm{s}^{-1}$ and a collision frequency of $\omega_{\mathrm{c}} = 1.12 \cdot 10^{14}\,\mathrm{s}^{-1}$ were used. In the pictured frequency range, the calculated values closely represent the experimental data for gold taken from [31] (circles). At higher frequencies the Drude model fails due to interband transitions that occur in the visible.

2.2 Metamaterials

Electromagnetic metamaterials are a class of man-made substances composed of nanoscopic, metallic building blocks whose shape and arrangement determine the properties of the material. Usually, metamaterials are realised by a repetition of the same, basic elements in a periodic arrangement. Due to the analogy with the internal charges of conventional materials, the constituent units are often called *artificial atoms*. As discussed in chapter 2.1.2, to facilitate the theoretical description of a conventional material the individual charges that compose it are substituted by a homogeneous medium with material parameters $\varepsilon_{\mathrm{r}}(\omega)$ and $\mu_{\mathrm{r}}(\omega)$. Analogously, in metamaterials, *effective material parameters* $\varepsilon_{\mathrm{r}}^{\mathrm{eff}}(\omega)$ and $\mu_{\mathrm{r}}^{\mathrm{eff}}(\omega)$ can, on a larger scale, replace the smaller-scale distribution of artificial atoms. This works because, due to their periodicity, the behaviour of the constituent elements and their interaction is sufficiently uniform. The new, averaged parameters are determined by the shape, size and spatial configuration of the artificial atoms and can be manipulated and tailored in a wide range. However, the artificial atoms and their spacing have to be significantly smaller than the averaging scale. Consequently, they have to be smaller than the wavelength associated with the frequency ω, for which the effective parameters are optimised.

Of special interest is the possibility to create elements with a magnetic response to externally applied electromagnetic fields, thus realising effective permeabilities which deviate from one at optical frequencies [32]. This effect is due to resonances of oscillating magnetic dipoles. With it, even negative μ_r^{eff} can be obtained [33]. If the effective permittivity is negative as well, this leads to a negative refractive index n [34–37]. That, in turn, gives rise to unusual results, such as a wave vector antiparallel to the Poynting vector, negative refraction [6], superlenses [7], and a reversed Cherenkov effect [8]. Moreover, metamaterials can be utilised to investigate the coupling of magnetic dipoles at optical frequencies [3]. In this thesis three types of metamaterials were examined: arrays made up of *split-ring resonators* [32], *double-wire* structures [4], and a so-called *fishnet* [5, 38]. These are introduced in the following sections.

2.2.1 Split-ring resonators

Split-ring resonators (SRRs) are among the most important and most widely used types of artificial atoms with magnetic response. Essentially, they are metallic rings with a gap, which were used in the gigahertz regime (200–2000 MHz) as early as 1981 by Hardy and Whitehead [39]. If viewed from above (see Fig. 2.2), an SRR can be compared to an antenna bent to almost form a spherical or square ring. Both the antenna and the SRR support several resonance modes:

In the fundamental mode of the split-ring resonator, the *magnetic mode* (Fig. 2.2 (a)), the conduction electrons move from one end of the SRR to the other, generating an oscillating electric dipole moment (blue arrow). They also form a circular current around the SRR, which causes a magnetic moment perpendicular to it (shown in red). In this mode, the SRR therefore exhibits a magnetic response. It can be excited resonantly by a magnetic field in the direction of the magnetic moment, but also, more strongly, by an electric field parallel to the electric moment. This means that split-ring resonators are bi-anisotropic. Therefore, a plane wave under normal incidence—*i.e.*, impinging perpendicularly to the plane of the SRR—can couple to the magnetic mode if it is polarised horizontally. The corresponding polarisation direction is indicated by the black double arrow on the right-hand side of Fig. 2.2 (a). The horizontal and vertical polarisation directions are the two orthogonal *eigenpolarisations* of the split-ring under

2.2 Metamaterials

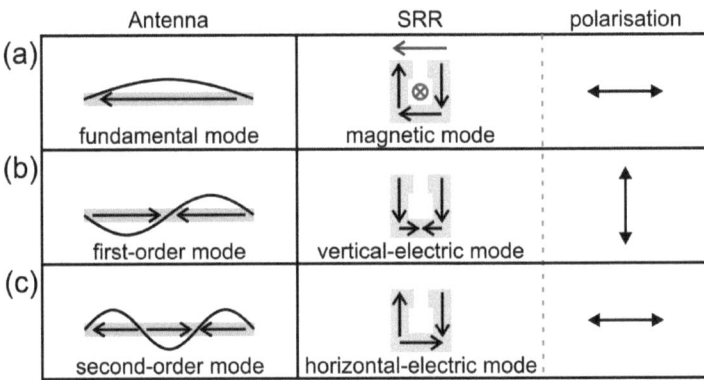

Fig. 2.2: Sketch of the first three modes of a linear antenna (left-hand side column) and a split-ring resonator (SRR, centre column). Snap-shots of the current distribution (black arcs), its direction (black arrows), as well as the electric and magnetic dipole moment of the fundamental mode (blue and red, respectively) are depicted. The arrows in the right-hand side column indicate the polarisation direction of a plane wave's electric field, with which the corresponding SRR mode can be excited. Adapted from [40].

normal incidence, as any polarisation other than the vertical one excites the electric dipole moment and thus couples to the horizontal polarisation. Depending on the dimensions of the SRR, negative μ_r^{eff} are possible in the magnetic mode.

The higher-frequency first-order mode features a current node in the centre of both antenna and SRR (Fig. 2.2 (b)). In the split-ring it has a purely vertical electric dipole moment, which allows excitation of the mode by a vertically polarised plane wave. As there are only higher-order magnetic moments, the magnetic response of this mode is comparatively small. The second-order mode, Fig. 2.2 (c), exhibits two current nodes and can be excited with a horizontal polarisation. Here, too, the magnetic response is weak.

Fig. 2.3 shows calculated far-field transmittance and reflectance spectra of an array of gold split-ring resonators under normal incidence illumination [41]. At the resonance frequency of the modes described above, the SRRs absorb and scatter the incident electromagnetic energy, causing a drop in the transmittance and a peak in the reflectance. As indicated in the insets, the magnetic and horizontal-electric modes are excited by the horizontally polarised wave while the vertical-electric mode is excited under vertical polarisation.

Obviously, the excitation of all modes strongly depends on the frequency of the impinging fields. Following [37], a simple model can be employed to estimate the frequency of the magnetic resonance: the SRR is described by its equivalent circuit, *i.e.*, an LC-circuit. The ring itself corresponds to an induction coil with a single winding, the gap is represented by the capacitor, and the impinging plane wave driving the resonance is substituted by a voltage source, as shown

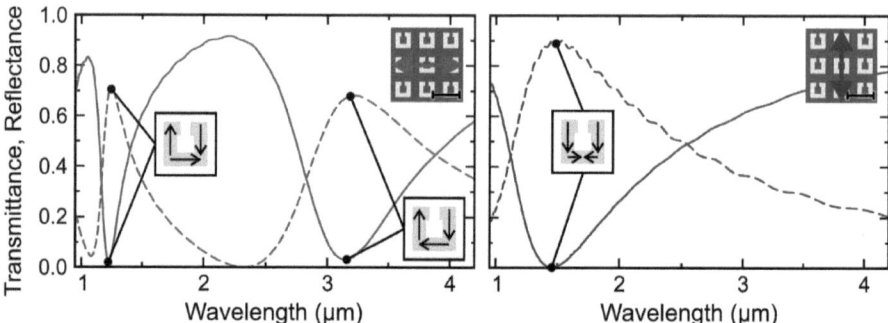

Fig. 2.3: Calculated far-field transmittance (continuous lines) and reflectance (dashed lines) spectra of an array of gold SRRs for illumination under normal incidence. The structure is shown in the upper right corners (the scale bar is 500 nm), along with the polarisation direction of the incident electric field. Insets show the current distributions for the corresponding modes. Adapted from [41], calculation parameters as in [32].

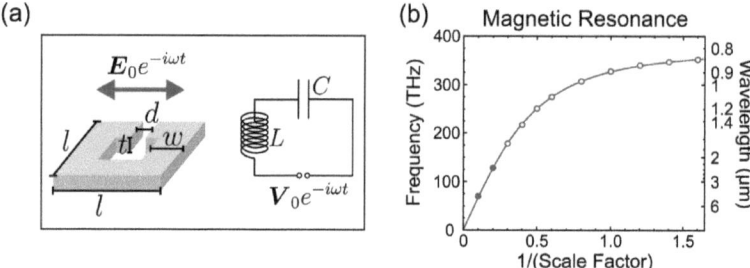

Fig. 2.4: (a) Sketch of an SRR with impinging plane wave (direction of the electric field shown in red), and the equivalent, externally driven LC-circuit. (b) Experimental results (symbols) and fit (continuous line) for the influence of size-scaling on the resonance frequency of SRRs, taken from [42].

in Fig. 2.4 (a). The damping in the SRR is neglected. For such an LC-circuit, the resonance frequency is

$$\omega = \frac{1}{\sqrt{LC}}. \qquad (2.15)$$

With the dimensions of the split-ring as designated in Fig. 2.4 (a), the capacitance C can be calculated with the formula for a plate capacitor, while the inductance L is approximated with the formula for a long cylindrical coil:

$$C = \varepsilon_0 \varepsilon_\mathrm{r} \frac{wt}{d} \quad \text{and} \quad L \approx \mu_0 \frac{l^2}{t}. \qquad (2.16)$$

2.2 Metamaterials

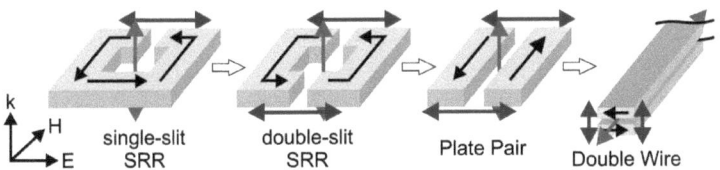

Fig. 2.5: Illustration of the transformation from a single-slit split-ring resonator to a double-wire configuration. Snap-shots of the currents are indicated by the black arrows, electric and magnetic moments by the blue and red double arrows, respectively.

Thus, the resonance frequency is given by

$$\omega \approx \frac{1}{l}\frac{c_0}{\sqrt{\varepsilon}}\sqrt{\frac{d}{w}}. \quad (2.17)$$

As can be seen from this formula, ω depends on the lateral dimensions of the split-ring, such that scaling down the SRR results in a higher resonance frequency and, correspondingly, a lower resonance wavelength. This conclusion holds true as long as the capacitance and inductance assumed above accurately represent the SRR. However, at higher frequencies the kinetic energy of the oscillating electrons cannot be neglected anymore. Also, ohmic losses and losses through radiation increase. Fig. 2.4 (b) shows experimental results for the position of the resonance on scaling the size of SRRs [42]. As expected, for large split-ring resonators the resonance frequency grows approximately linearly with decreasing SRR size. For small resonators, however, formula 2.17 ceases to be valid and the resonance frequency stagnates, so that by scaling alone resonance wavelengths below roughly $\lambda \approx 800$ nm cannot be achieved. Moreover, negative μ_r^{eff} are obtained only down to $\lambda \approx 1.15$ µm [43].

2.2.2 Double-wires

Magnetic resonances at shorter wavelengths, particularly in the visible spectrum, are possible if the SRR design is altered instead of merely scaled. Fig. 2.5 shows the transformation of a single-slit SRR, first to a double-slit split-ring, then to a plate pair, and finally to a double-wire configuration with infinitely long wires. (Note, that in this thesis the term "double-wire" is employed only for infinitely long wires. With a length of 100 µm, the wires used in chapter 5 sufficiently approximate this infinite length.)

In the single-slit SRR the fundamental mode generates the magnetic dipole moment (red double arrow) by its circular current. This circular current becomes an antisymmetric oscillation in the other structures of Fig. 2.5, also producing strong magnetic dipole moments. Thus, the plate capacitor at the gap of the split ring is, in the double-wires, transferred to the termination points of the currents, the edges of the two wires, where charges accumulate opposite to each

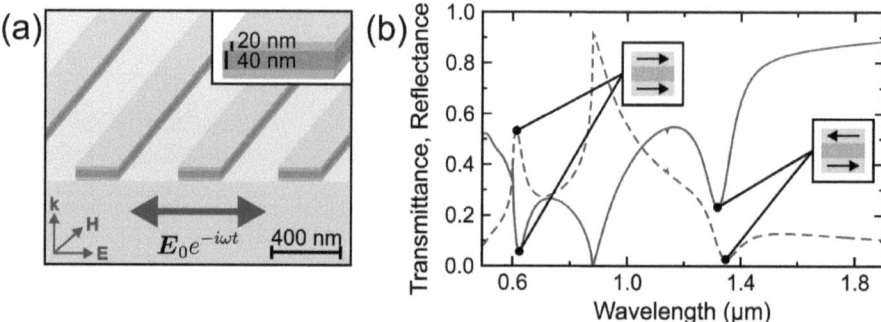

Fig. 2.6: (a) Array of gold double-wires with a spacer of MgF$_2$ (green). The width of the wires is 250 nm, the pitch is 600 nm. (b) Calculated far-field transmittance (continuous line) and reflectance (dashed line) spectra of the structure shown in (a), with illumination at normal incidence from below, as indicated by the red double arrow.

other. Therefore, while the magnetic field is concentrated in the space between the wires, the electric field is strongest between the edges. Moreover, the opposing electric moments cancel at some distance from the structure, so that the resulting far-field electric moment vanishes. The described mode can be interpreted as the lower-energy, antisymmetric eigenmode resulting from the coupling of the fundamental electric dipole resonances of the individual wires. In the higher, symmetric eigenmode the currents in the two wires oscillate in phase. This mode exhibits only an electric dipole moment across the wires and no magnetic dipole moment. Both modes of the double-wires can be excited under normal incidence, with the electric field polarised perpendicularly to the wires. If excited by light polarised parallel to the wires, no comparable resonances occur and the wires act as a *diluted metal*. Fig. 2.6 (a) shows the sketch of a double-wire structure, and Fig. 2.6 (b) calculated far-field transmittance (continuous line) and reflectance (dashed line) spectra. The calculations were performed with the scattering-matrix algorithm described in section 2.3.2. In the insets the current distributions for the symmetric and antisymmetric mode are depicted. The positions of the modes are determined by the width and thickness of the wires, as well as by the distance between the two metal layers and the spacer material. Wide but thin wires, with a thin, high-permittivity spacer material have the smallest resonance frequencies, while more circular wire cross-sections cause higher frequencies.

As can be seen in Fig. 2.6 (b), at a wavelength of approximately 900 nm the *Wood anomaly* causes a pronounced drop in transmittance. This effect was discovered by Robert Wood in 1902 [44] and explained theoretically by Lord Rayleigh [45] five years later. It is caused by the periodicity of the double-wires: inside any periodic structure the in-plane components k_x and k_y of the wave vector of a wave are only conserved to within multiples of the reciprocal lattice

2.2 Metamaterials

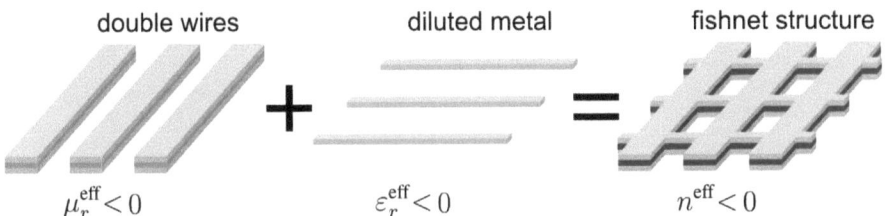

Fig. 2.7: Illustration of the composition of the fishnet structure. Double-wires with negative μ are combined with a diluted metal (metal strips with negative ε) to give a structure with negative n that resembles a fishnet. Adapted from [5].

vectors g_x and g_y. Adding g to k results in a reduction of k_z, because the magnitude of the wave vector is conserved. For those wavelengths or angles of incidence at which k_z reaches zero, the diffracted wave propagates parallel to the structure and the Wood anomaly occurs. At such angles and wavelengths the structure cannot be described by effective material parameters.

An interesting effect can be observed if a double-wire array is placed on top of a waveguide, such as a dielectric layer. In this configuration the wires on top of the waveguide impose their periodicity on the guided modes. When this periodicity is scaled, the Bragg resonance of the waveguide shows an avoided crossing with the magnetic resonance of the wires [4]. An analogous effect is found for the similar single-wire structures on waveguides. These consist of only the bottom layer of double-wires and, therefore, possess no magnetic resonance. However, their fundamental resonance also exhibits an avoided crossing with the waveguide mode [46]. Other metamaterial designs related to the double-wires are *cut-wire pairs* and *plate pairs* [47]. They are obtained by shortening the virtually infinitely long double-wires into rectangular and square plates, respectively. As is the case with a more circular wire cross-section, a less pronounced disparity between length and width of the wires results in a shorter resonance wavelength.

In chapter 5 we present spectrally resolved measurements of the near-field distribution of double-wire structures along with calculated distributions. Calculations of single-wire structures are shown in chapter 4.3.

2.2.3 Fishnet structure

In order to achieve a negative refractive index at optical frequencies, without incurring severe losses, both the effective permittivity and the effective permeability should be negative ("double-negative" materials). A design proposed as well as implemented by Zhang *et al.* [5,38] is the *fishnet structure*. As sketched in Fig. 2.7, it consists of an array of double-wires, which provide the magnetic response, crossed by a second array of narrower wires. These narrow wires are parallel to the polarisation of the incident light, so that they serve as a *diluted metal*.

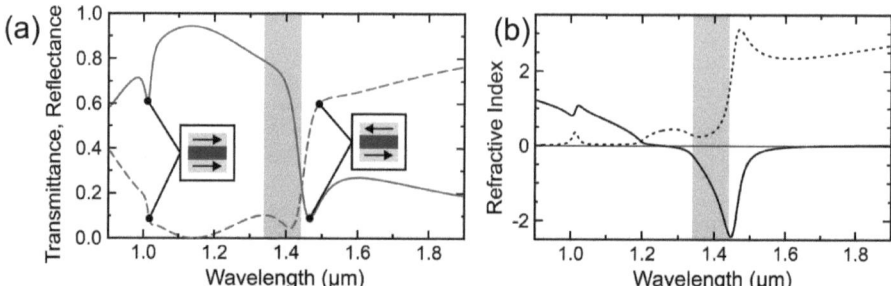

Fig. 2.8: (a) Exemplary calculated far-field transmittance (continuous line) and reflectance spectra (dashed line) of a fishnet structure under normal incidence, with the polarisation parallel to the wires of the diluted metal. (b) Real (continuous line) and imaginary part (dashed line) of the corresponding refractive index. Adapted from [49].

To facilitate the fabrication, the diluted metal is produced with the same three layers as are the double-wires: one gold layer each at the bottom and the top, and a dielectric spacer layer between them. The resulting design can be manufactured small enough for the negative index to reach visible wavelengths [48].

Fig. 2.8 shows (a) the calculated far-field transmittance (continuous line) and reflectance (dashed line) spectra, as well as (b) the real part (continuous line) and the imaginary part (dashed line) of the refractive index of an exemplary fishnet structure (adapted from [49]). The insets illustrate the resonance modes in the double-wires. Due to its composition from double-wires and a diluted metal, the fishnet structure behaves like a diluted metal for wavelengths far longer than the resonance of the double-wires. Here, the transmittance is low, the reflectance is high, and the real part of the refractive index lies just below zero. Like in a typical metal, however, losses are high due to the high imaginary part of n, preventing efficient use of the negative index. In contrast, close to the magnetic resonance below 1.5 μm, the permeability drops below zero, thus allowing a negative real part of the refractive index with lower losses. Correspondingly, the transmittance increases. Further from the antisymmetric resonance the negative permittivity and permeability are lost and the refractive index rises above zero. Here, the fishnet behaves like a lossy dielectric. At about 1 μm the symmetric resonance can be seen as a dip in transmittance. Clearly, that spectral region is of the greatest interest, where the refractive index is notably below zero, while the losses are still acceptable. The corresponding range of wavelengths is highlighted in grey in Fig. 2.8.

Amplitude and phase sensitive near-field measurements of the field distribution on top of a fishnet structure are presented in chapter 6.

2.2 Metamaterials

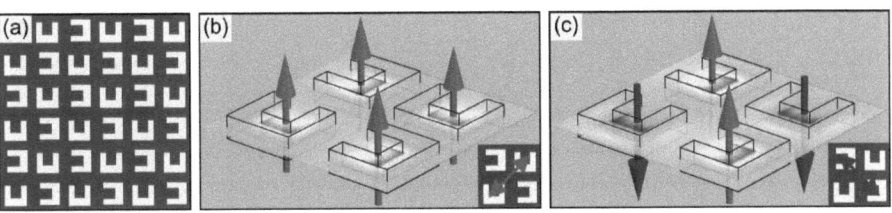

Fig. 2.9: (a) Sketch of a low-symmetry split-ring-resonator array. Under normal incidence of diagonally polarised light (see insets) the magnetic field component parallel to the propagation direction reveals the parallel (b) and antiparallel (c) magnetic moments. Taken from [3].

2.2.4 Magnetic interactions between artificial atoms

In addition to negative refractive indices, interactions taking place between the magnetic moments of neighbouring artificial atoms are of great interest [50–54]. Obviously, for the investigation of such magnetic couplings, metamaterial designs with little or no electric interactions between the elements are suited best, because the coupling effects found in these structures can be attributed mainly to the magnetic interactions. Planar, low-symmetry split-ring-resonator arrays, proposed and studied in the far-field by Decker *et al.* [3, 41], meet this requirement. They consist of an array of individual SRRs which are alternately rotated by 0 and 90 degrees, as shown in Fig. 2.9 (a). In this configuration every resonator is directly surrounded only by split-rings with an orthogonal orientation. This means that the electric dipole moments of any pair of neighbouring SRRs are perpendicular. Thus, no next-neighbour electric dipole–dipole interactions take place. Contrarily, the magnetic moments of the resonators are all parallel and can couple to each other.

To understand the effect of this coupling we consider how the array would behave if no magnetic interactions occurred. As noted in chapter 2.2.1, under normal-incidence the eigenpolarisations of split-rings are horizontal and vertical. For the resonators rotated by 90 degrees, that is, with their gap pointing to the left, they are vertical and horizontal. Thus, without interaction between the SRRs, vertically polarised light would couple only to the fundamental mode of the rotated resonators and horizontally polarised light only to the non-rotated ones. Therefore, the eigenpolarisations for the combined array would also be horizontal and vertical and no conversion between them could be observed.

However, with interactions between the split-ring resonators, energy is transferred from the excited resonators to the ones oriented orthogonally. Thus, some of the impinging light is converted to the perpendicular polarisation. The strength of that conversion is the same regardless of which subset of resonators is excited. In the formalism of the Jones calculus the Jones matrix

of the SRR array therefore reads
$$\begin{pmatrix} A & \pm B \\ \pm B & A \end{pmatrix}. \qquad (2.18)$$
The eigenvectors of this matrix are the diagonal polarisation directions, $+45°$ and $-45°$. Figs. 2.9 (b) and (c) visualise this in two snap-shots of the magnetic field component parallel to the propagation direction. Under illumination by $+45°$, shown in (b), the magnetic moments of the rotated and the original resonators are parallel. This stays true for any point in time (except when they are zero). Due to the analogy with spins in solids, this mode is named "ferromagnetic". For the eigenpolarisation of $-45°$, shown in (c), neighbouring moments always, except when zero, point in opposite directions ("antiferromagnetic" mode). As would be expected, the parallel magnetic moments are associated with a higher energy than the antiparallel ones. Consequently, the resonance frequency for the ferromagnetic eigenpolarisation is shifted to a higher frequency with respect to that of an individual split-ring resonator, while the resonance of the antiferromagnetic eigenpolarisation is shifted to lower frequencies.

In the experiments shown in Fig. 2.10, such a frequency shift can be observed clearly. An SEM image of the examined structure is displayed in Fig. 2.10 (a), far-field intensity transmittance spectra under normal incidence in Fig. 2.10 (b). The polarisation directions of the illuminating light are colour coded as defined in the insets. In the top row of Fig. 2.10 (b) we show the transmittance polarised parallel to the incident polarisation, $T_\parallel$. As guides to the eye, the dashed lines indicate the two different resonance frequencies of the diagonal polarisation directions, 232 THz and 240 THz (1290 nm and 1250 nm). The bottom row of Fig. 2.10 (b) gives the conversion spectra, $i.e.$, the transmittance polarised perpendicularly to the incident light, $T_\perp$. When illuminated with vertically or horizontally polarised light, the sample converts more than one percent of that light into the orthogonal polarisation, whereas there is no detectable conversion between the diagonal polarisation directions, the eigenpolarisations.

Measurements of the near-field distributions of low-symmetry split-ring-resonator arrays are presented in chapter 7.

2.3 Calculation tools

Maxwell's equations render it possible to check the results of experiments against theory. However, the permittivity distributions of most metamaterials are far too complex to solve the ensuing equations analytically. Therefore, computational approaches are used. Two algorithms were employed in this work and are outlined in this chapter: the *Finite-Integration Technique* and the *Scattering-Matrix Method*.

2.3 Calculation tools

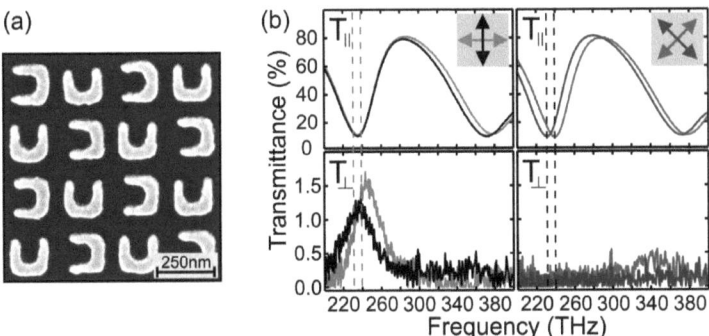

Fig. 2.10: (a) SEM image of a low-symmetry split-ring-resonator array sample. (b) Normal-incidence spectra from this sample (incident polarisation directions defined in the insets). Far-field transmittance polarised parallel (top row) and perpendicular (bottom row) to the polarisation of the incident light. The spectra show significant conversion for vertical and horizontal polarisations, as well as a resonance frequency splitting for the diagonal eigenpolarisations of the structure, where no conversion is observed. Taken from [41].

2.3.1 Finite-integration technique

Almost all of the near-field distribution calculations, as well as the far-field transmittance and reflectance computations in this work were performed with the commercially available simulation tool "CST MICROWAVE STUDIO" by CST AG. The algorithm employed by it is the Finite-Integration Technique (FIT), which can be implemented both for time and for frequency domain calculations. As only the former were used in this work, the following description is limited to the time domain algorithm.

Like the FDTD (*Finite-Difference Time-Domain*) method, which is the most commonly used time domain approach, the FIT is based on a spatial discretisation scheme proposed by Kane Yee in 1966 [55]: a grid is laid over the computational domain, partitioning it into tiny elements, for instance cubes or tetrahedra—the *Yee cells*. Then, the components of the electric and magnetic fields ($\boldsymbol{E}$ and $\boldsymbol{H}$) are sampled at different positions, such that the electric field is known at the centre of each edge of all Yee cells, while the magnetic field is known at the centre of each face. Importantly, the electric and magnetic fields are also sampled at slightly different times t and $t + \frac{1}{2}\Delta t$, respectively.

We stress that in contrast to the more common FDTD method, which utilises Maxwell's equations in their differential form, the FIT makes use of the integral form (see right-hand side of Eqs. (2.1)). As can be seen there, the time derivative of $\boldsymbol{E}$ depends on the integral over $\boldsymbol{H}$ in space. Hence, for a known field distribution of $\boldsymbol{E}(t)$ and $\boldsymbol{H}(t + \frac{1}{2}\Delta t)$ the distribution of $\boldsymbol{E}(t+\Delta t)$ can be computed. Likewise, the time derivative of $\boldsymbol{H}$ depends on the integral over $\boldsymbol{E}$, so $H(t+\frac{3}{2}\Delta t)$ can be calculated from $E(t+\Delta t)$ and $H(t+\frac{1}{2}\Delta t)$. Following that, $E(t+2\Delta t)$ is determined and so forth. In this so-called *leap-frog* process, successive and alternating updates are made on the fields. This is continued until either a steady state is reached or the fields have decayed within the entire modelled area.

Instead of beginning with a fixed field distribution, the calculation can be started with a wave pulse impinging on the edge of the computational domain or coming from a source within it. Then, in a single run, the behaviour of a sample can be computed for the entire range of frequencies contained in that wave pulse.

2.3.2 Scattering-Matrix Method

The scattering-matrix code used in this work was implemented in Matlab (by The MathWorks, Inc.) by Stefan Linden. It is designed for structures made up of layers that are homogeneous in one in-plane direction and periodical in the other. In general, periodicity in both in-plane directions suffices for scattering-matrix calculations. An extensive and detailed discussion of the algorithm can be found in [56], a short overview is given here.

Scattering-matrix calculations are executed to find a relation between the complex fields impinging on a planar, periodical structure, and the fields reflected and transmitted by it. To this end, the sample is partitioned into a stack of N layers (see Fig. 2.11 (a)), each of which is periodic in x and y, and homogeneous in the z-direction. Then, the incident fields E and H are expanded into Bragg orders of plane waves. This expansion is truncated, making it the only approximation of the scattering-matrix method. The resulting plane-wave coefficients are combined into a single vector of complex amplitudes (simplifyingly depicted as black arrows in Fig. 2.11). This is done for every layer, starting with the waves travelling in the positive z-direction (but not necessarily parallel to the z-direction) towards the sample, as well as with those incident from the opposite side going towards $-z$. For these two this leads to the vectors $\mathcal{A}_0^+$ and $\mathcal{A}_{N+1}^-$, respectively. The superscripts on the coefficient vectors, $+$ and $-$, denote the propagation direction, while the subscripts indicate the number of the layer the waves are propagating in: 1 to N are the layers of the structure in consecutive order, 0 is the space before the structure (*e.g.*, a substrate), and $N+1$ designates the space behind it (*e.g.*, vacuum).

For every layer n, a transfer matrix $\mathbb{T}_P$ is calculated to propagate the Bragg amplitudes from the front face of the layer at $z - L$ to the end face at z, see Fig. 2.11 (b). For every boundary, another transfer matrix, $\mathbb{T}_B$, is computed. It implements the continuity equations to link the fields on both sides of the boundary between the layers n and $n+1$. Thus, $\mathbb{T}_P$ and $\mathbb{T}_B$ connect the coefficient vectors across the layers:

$$\begin{pmatrix} \mathcal{A}_n^+(z) \\ \mathcal{A}_n^-(z) \end{pmatrix} = \mathbb{T}_P \cdot \begin{pmatrix} \mathcal{A}_n^+(z-L) \\ \mathcal{A}_n^-(z-L) \end{pmatrix} \quad \text{and} \quad \begin{pmatrix} \mathcal{A}_n^+(z) \\ \mathcal{A}_n^-(z) \end{pmatrix} = \mathbb{T}_B \cdot \begin{pmatrix} \mathcal{A}_{n+1}^+(z) \\ \mathcal{A}_{n+1}^-(z) \end{pmatrix} \quad (2.19)$$

2.4 Near-field microscopy

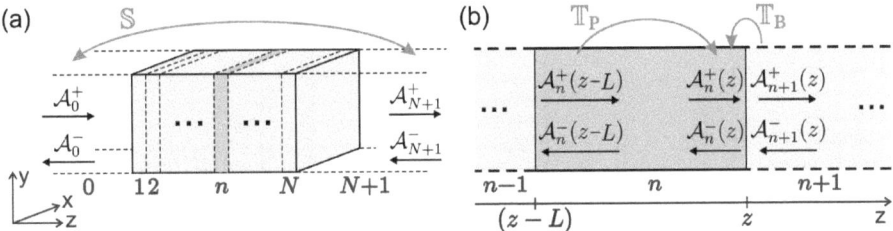

Fig. 2.11: (a) Illustration of a photonic structure partitioned into N layers. The scattering matrix $\mathbb{S}$ links the amplitude-coefficient vectors of the incident fields ($\mathcal{A}_0^+$ and $\mathcal{A}_{N+1}^-$) with those leaving the structure ($\mathcal{A}_{N+1}^+$ and $\mathcal{A}_0^-$). (b) Detail of the n-th layer. The transfer matrices $\mathbb{T}_\mathrm{P}$ and $\mathbb{T}_\mathrm{B}$ are used to propagate the amplitudes across a single layer and a single boundary, respectively. Based on [56].

From these equations, a new matrix can be derived relating the fields incident on layer n with the fields leaving it. This is the scattering matrix for the n-th layer. Step by step, the scattering-matrix algorithm incorporates the adjacent boundary and transfer matrices into the scattering matrix, until the entire sample is comprised in it. In the end, one simple equation characterises the electromagnetic properties of the structure:

$$\boxed{\vec{\mathcal{A}}_\mathrm{out} = \mathbb{S} \cdot \vec{\mathcal{A}}_\mathrm{in}} \quad \text{with} \quad \vec{\mathcal{A}}_\mathrm{in} = \begin{pmatrix} \mathcal{A}_0^+ \\ \mathcal{A}_{N+1}^- \end{pmatrix} \quad \text{and} \quad \vec{\mathcal{A}}_\mathrm{out} = \begin{pmatrix} \mathcal{A}_{N+1}^+ \\ \mathcal{A}_0^- \end{pmatrix}. \qquad (2.20)$$

2.4 Near-field microscopy

While theoretical and numerical approaches allow us to calculate the electromagnetic fields in and around structures of interest in high detail, the desired experimental determination proves quite difficult. This is due to the limited resolution of classical microscopy. For our investigation of the metamaterials introduced in chapter 2.2 we therefore used a near-field microscope.

In the following section we recapitulate the classical resolution limit, then give a theoretical analysis of the higher resolution attainable in near-field measurements. We end with an overview over the experimental implementation of near-field microscope setups.

2.4.1 Resolution limit

It is well known that the resolution that can be obtained with conventional microscopes is fundamentally limited. At the end of the nineteenth century Lord Rayleigh and Ernst Abbe independently quantified this limit [57, 58]. Following Abbe, two point sources can still be

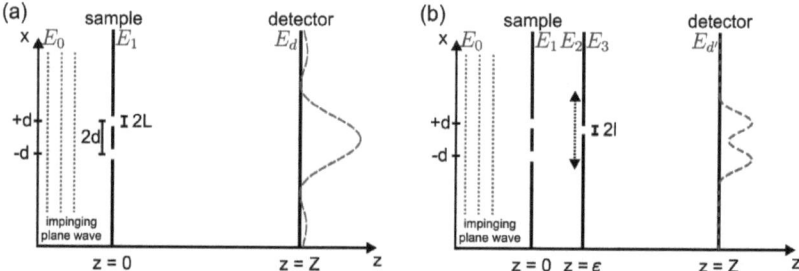

Fig. 2.12: Illustration (not to scale) of the higher resolution attainable in near-field microscopy: a plane wave impinges on a double-slit sample, generating (a) a diffraction limited image on the screen and (b) a more highly resolved image due to a movable aperture placed closely behind the sample. Based on [60].

separated if their distance is at least

$$\Delta x \geq \frac{0{,}61 \cdot \lambda_0}{n\sin(\theta)}, \quad (2.21)$$

where λ_0 is the vacuum wavelength of the light used, n the refractive index and θ half the aperture angle of the microscope's objective. By choosing large objectives close to the sample and employing immersion oil with a large refractive index, the *numerical aperture* $NA = n\sin(\theta)$ can be increased. Nevertheless, a resolution limit of approximately $\Delta x \approx \lambda_0/2$ remains, which restricts us to about 750 nm for a wavelength of 1510 nm.

Distinctly higher resolution is possible, for example with *Stimulated Emission Depletion microscopy* (STED) and *Photo-activated Localisation Microscopy* (PALM or STORM for *Stochastic Optical Reconstruction Microscopy*), all of which, however, require fluorescent or photoactive samples (for an overview see [59]). Near-field microscopy, however, does not share this requirement and thus can be used for our investigation of metamaterials.

2.4.2 Resolution enhancement in the near-field

The following section covers a theoretical analysis of the resolution achievable with a near-field microscope. It closely follows a derivation by Vigoureux *et al.* [60].

We consider a sample consisting of two nanoscopic openings in an otherwise opaque film at $z = 0$. Both openings are $2L$ wide and are located at $x = +d$ and $-d$ (see Fig. 2.12 (a)). A plane wave $E_0(x, z)$ impinges from $-z$, hence the field directly behind the film is the product of E_0 and an aperture function T, which equals 1 in the openings and 0 elsewhere. In the spatial

2.4 Near-field microscopy

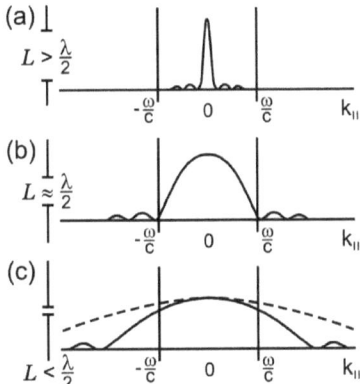

Fig. 2.13: Spatial frequency spectrum of three differently sized objects: (a) larger than half the wavelength λ, (b) comparable to it, and (c) substantially smaller. Adapted from [61].

frequency domain this field can be written as

$$E_1(k_x, z=0) = 4E_0 \cos k_x d \, \frac{\sin k_x L}{k_x}, \tag{2.22}$$

with the wave number or *spatial frequency* denoted as k_x. A Fourier transform produces the field distribution on a detector at a distance Z from the film:

$$E_d(x, z=Z) = \frac{1}{2\pi} \int_{-\infty}^{+\infty} e^{-ik_x x} \, E_1(k_x, z=0) \, e^{-i\sqrt{k^2-k_x^2}\,Z} \, dk_x, \tag{2.23}$$

where we have replaced k_z^2 by using $k_z^2 = k^2 - k_x^2$. In this integral k_x takes on all values from $-\infty$ to $+\infty$ while $k = \omega/c$ is constant. Therefore, k_z becomes imaginary for waves with $|k_x| > \omega/c$. Consequently, instead of oscillating, such waves will decay exponentially with distance. To this they owe the name *evanescent fields* or *near-fields*. The larger k_x grows, the larger $|k_z|$ will be, causing a stronger decay. Thus, on a detector placed at some distance from the sample (at $z = Z$, see Fig. 2.12 (a)), the contribution of these waves vanishes and we can approximate:

$$E_d(x, z=Z) \approx \frac{1}{2\pi} \int_{-\frac{\omega}{c}}^{+\frac{\omega}{c}} e^{-ik_x x} \, E_1(k_x, z=0) \, e^{-i\sqrt{k^2-k_x^2}\,Z} \, dk_x. \tag{2.24}$$

This is the field a conventional microscope can detect. It includes only the *propagating waves* or *far-fields* with spatial frequencies $|k_x| \leq \omega/c$.

As can be seen from Fig. 2.13 (a), almost all of the spatial frequencies of an object lie within the detectable range from $-\omega/c$ to $+\omega/c$, provided that the object is substantially larger than half the wavelength of the utilised light (or of comparable size, as in 2.13 (b)). Such objects can thus be reproduced with reasonably high accuracy. Of notably smaller objects, as in 2.13 (c), most of the spatial frequencies are lost and with them a large part of the information. The spectra of an object of this size (continuous line) and an even smaller one (dashed line) are so similar, that resulting images of them are impossible to differentiate. Therefore, to resolve details below $\lambda/2$, higher spatial frequencies have to be picked up.

In order to make this possible we consider an extended setup, depicted in Fig. 2.12 (b): an additional screen with an aperture is placed as a near-field probe at a distance ε behind the double-slit sample. The opening has a width of $2l$ and its centre lies at an arbitrary point $x = X$. According to Eq. (2.23), the field directly in front of this screen is given by

$$E_2(x, z = \varepsilon) = \frac{1}{2\pi} \int_{-\infty}^{+\infty} e^{-ik_x x} E_1(k_x, z = 0) e^{-i\sqrt{k^2 - k_x^2} \varepsilon} dk_x. \tag{2.25}$$

As above, multiplication with an aperture function yields the field behind the probe. In the frequency domain this reads:

$$E_3(k_x, z = \varepsilon) = \frac{1}{2\pi} \int_{-\infty}^{+\infty} E_1(k_x', z = 0) e^{-i\sqrt{k^2 - k_x'^2} \varepsilon} 2\frac{\sin(k_x - k_x')l}{k_x - k_x'} e^{i(k_x - k_x')X} dk_x'. \tag{2.26}$$

Again, the field distribution on the detector at $z = Z$ is obtained by a Fourier transform and subsequent discarding of the evanescent components:

$$E_{d'}(x, z = Z) \approx \frac{1}{(2\pi)^2} \int_{-\frac{\omega}{c}}^{+\frac{\omega}{c}} e^{-ik_x x} e^{-i\sqrt{k^2 - k_x^2}(Z - \varepsilon)}$$
$$\times \int_{-\infty}^{+\infty} 4E_0 \cos k_x' d \frac{\sin k_x' L}{k_x'} e^{-i\sqrt{k^2 - k_x'^2}\varepsilon} 2\frac{\sin(k_x - k_x')l}{k_x - k_x'} e^{i(k_x - k_x')X} dk_x' dk_x. \tag{2.27}$$

The decisive difference between this result for the near-field setup and the earlier result from the conventional setup (Eq. (2.24)) lies in the integration over

$$\frac{\sin(k_x - k_x')l}{k_x - k_x'} e^{i(k_x - k_x')X} \tag{2.28}$$

in the inner integral, which causes a shift from k_x to $k_x - k_x'$. Due to this, the collected spatial frequencies are shifted by k_x' with respect to the earlier, conventional setup. They now include formerly evanescent modes together with the associated information. This is the key result of this analysis: the near-field probe in the extended setup causes evanescent modes to reach the detector and thus enables a far higher resolution than a conventional microscope. A further inspection of Eq. (2.27)—especially of the terms with ε in the exponential function—shows, that this higher resolution strongly depends on the distance between probe and sample, ε. Finally, note that for large probe apertures, $l \to +\infty$, the first term of Eq. (2.28) converges

2.4 Near-field microscopy

towards the delta function and the field of Eq. (2.27) converges towards the field detected by the conventional setup, Eq. (2.24). Thus, the size of the probe aperture also crucially influences the resolution attainable with the near-field microscope.

To summarise our theoretical analysis: the smaller the size of the probe aperture l and the smaller the distance between probe and sample ε, the higher the resolution of a near-field microscope. How such a setup can be realised in practice is outlined in the following section.

2.4.3 Experimental implementation of near-field measurements

In this section we discuss different concepts for near-field experiments and distance control. The methods employed for *scanning near-field optical microscopy* (SNOM) are extremely varied. Generally, as in the previous section, a nanoscopic object is brought close to the sample. There, it converts the near-fields into propagating fields, which can be detected at a more convenient distance. Thus, near-field microscopes can be classified by the type of probe used, that is, *aperture probes* or *apertureless probes*. These are discussed in the following sections, along with the different modes of distance control. Further detailed overviews can be found in [62] and [63].

Apertureless probes

In near-field microscopy with apertureless probes—also called *scattering type* near-field microscopy—both sample and probe can be illuminated by an external source (see Fig. 2.14 "scattering") [20]. Probes can, for instance, be cantilevers from scanning force microscopy. They need a nanoscopically small tip to convert the incoming far-field to near-fields concentrated on the probe apex. The probe is raster scanned in close proximity to the sample and the reflected light is detected. The high background signal due to the external illumination can be suppressed with a heterodyne detection scheme [66]. Another method is illumination by total internal reflection ("TIR"), where the external light is reflected at the base of the sample and only near-fields reach the probe. With apertureless probes resolutions as high as 1–10 nm at a wavelength of 635 nm, as well as $\lambda/1\,000\,000$ in the microwave regime [67, 68] have been realised.

Aperture probes

In general, aperture probes consist of a nanoscopic opening in an otherwise opaque film. The first near-field probe proposed by Synge in 1928 was such an aperture in a flat metal screen [11]. Today, the most commonly used type of aperture probes is an optical fibre with a sharp

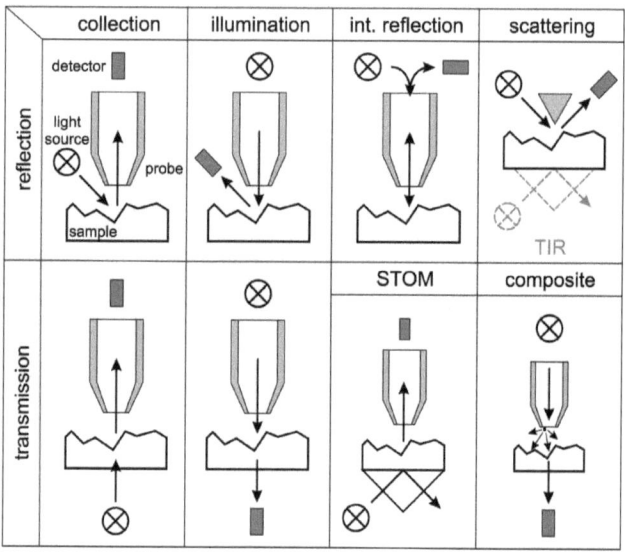

Fig. 2.14: Various probe concepts of near-field microscopy. Adapted from [64,65].

tip [16] produced either by heating and pulling [16,17] or by etching [18,69]. It is metallised and the aperture is located at the apex of its tip. Also used are tetrahedral probes [70,71].

As with apertureless probes the sample can be illuminated externally. Depending on the position of the light source the probe collects the reflected or the transmitted light (see Fig. 2.14 "collection") and conveys it to the detector. Here too, illumination by total internal reflection is used (*scanning tunnelling optical microscope*, see Fig. 2.14 "STOM"). The method employed in this work is a transmission/collection scheme which is detailed in chapter 3.2. In contrast to apertureless probes, the aperture probe can itself constitute the source, illuminating a nanoscopic part of the sample underneath the aperture. The fluorescence or reflection from the sample can then be measured by an external detector (Fig. 2.14 "illumination") or be picked up by the probe and return through the fibre (*internal reflection*, see Fig. 2.14 "int. reflection"). In order to increase the signal and/or augment resolution, the near-field from the aperture can be further concentrated with the aid of nanoscopic particles, for instance molecules or metal needles (Fig. 2.14 "composite").

The greatest advantage of these techniques lies in the strong suppression of background light due to the metallic film covering the probes. Typically, they provide a resolution of roughly ten percent of the wavelength utilised, but can achieve resolutions of a few percent [72]. Due to the cut-off at the tapered tip, the transmittance of aperture probes is usually noticeably below $T \approx 10^{-4}$ [62]. It decreases with the sixth power of aperture radius (see [12,13]) and thus

2.4 Near-field microscopy

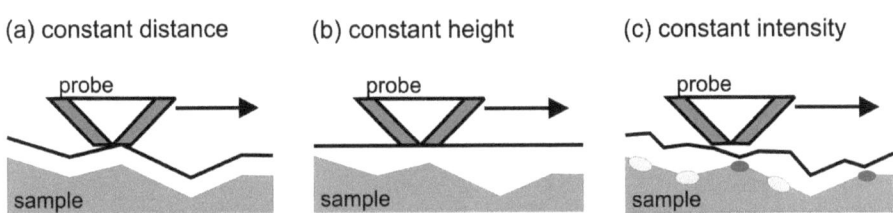

(a) constant distance (b) constant height (c) constant intensity

Fig. 2.15: Sketch of scanning modes: the probe can be scanned at a constant distance from the sample (a), (b) maintained at a fixed height or (c) traced along a line of constant intensity.

presents a limiting factor for smaller apertures and greater resolution. Generally, probes with larger opening angles feature a shorter cut-off area and higher transmittance.

Modes of distance control

Irrespective of probe type the near-field is detected only on the part of the sample closest to the probe. Therefore, in order to obtain a complete image, the probe has to be raster scanned across the surface of the sample. This can be done at a *constant distance* between probe and sample, following surface asperities (see Fig. 2.15 (a)). The distance is measured by means of strongly separation-dependent interactions such as shear forces. If the probes are small enough, these can be detected using cantilevers from atomic force microscopy. For larger probes, piezoelectric tuning forks are used. Alternatively, the tunnel effect can be made use of. Here, the probe–sample separation is kept invariable by maintaining the tunnel current constant. Scanning at a fixed distance facilitates the compensation of drifts in probe–sample separation and also permits the recording of the sample's topography in addition to the optical signal. However, measurement artefacts can arise due to the changes in absolute height [73]. These can be avoided by scanning at a predetermined height (see Fig. 2.15 (b), *constant height*), at the cost of losing the topographical data. A third method is the *constant-intensity* mode (Fig. 2.15 (c)), which is based on keeping the measured near-field signal intensity constant, retracting the probe from the sample at bright spots and approaching over dark areas. However, unless the surface of the sample is entirely flat, height changes of the probe in this mode can cause additional difficulties for the interpretation of the obtained images.

3 Experimental methods

In this chapter we present the experimental methods used in this thesis. Firstly, the processes involved in the fabrication of the aperture probes are described (chapter 3.1). Secondly, we summarise the setup of our scanning near-field microscope (chapter 3.2). Then, the fabrication of the metamaterial samples is sketched and it is shown how they are characterised in the far-field (chapters 3.3 and 3.4). Lastly, we introduce the operating mode of the phase- and polarisation-sensitive near-field microscope of the group of Prof. Kuipers which has been used to conduct measurements on double-wires and fishnet structures (chapter 3.5).

3.1 Probe fabrication

The transmission characteristics of the optical fibre probe are decisive for the resolution and signal intensity of the near-field microscope. Fabricating good probes therefore plays an important role in the success of measurements. The fabrication consists of an etching process, the metallisation and formation of an aperture and, finally, the mounting on the piezoelectric sensor used for distance control.

3.1.1 Chemical etching process

As we have seen in chapter 2.4.2, the aperture of a near-field probe has to be both very small and very close to the sample in order to produce a good resolution. If the aperture sits in a large, flat end face of an optical fibre, then a small angle between this plateau and the sample implies that the aperture ceases to be the point of the probe closest to the sample. Thus, it results in a limit on the proximity achievable. Therefore, the fibre end has to be tapered. A cheap and fast method to taper optical fibres is a combination of heating and pulling. However, with this method only fibres with small opening angles can be obtained, which suffer high losses due to waveguide cut-off [62]. Higher opening angles can be achieved by means of chemical etching. The method we use is a two-step chemical etching process sketched in Fig. 3.1. It follows a technique of Mononobe and Ohtsu [18], adapted by Adelmann in his Diplom thesis, where it is described in great detail [74].

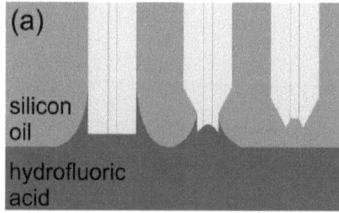 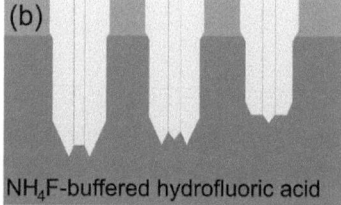

Fig. 3.1: Sketch (not to scale) of the two-step chemical etching process employed to taper optical fibres. (a) A shrinking meniscus of hydrofluoric acid (HF) generates the primary cone. (b) Ammonium fluoride buffered HF etches back the cladding, and the secondary cone—the probe's tip—emerges.

We use single-mode fibres for wavelengths down to 980 nm (Corning HI 980) which are cleaned and cleaved before the first etching step. Then the fibre end is dipped into 38 % v/v (volume per volume) hydrofluoric acid (HF). Monitored by a microscope, it is immediately pulled back slightly above the surface such that a short meniscus develops which links the fibre to the acid. As the fibre is corroded, its diameter decreases and the meniscus diminutes (see Fig. 3.1 (a)). Hence, with time, less and less of the fibre is etched and a cone—the *primary cone*—is generated. After approximately an hour the meniscus drops away from the fibre and the process terminates itself. The properties of the meniscus and of the cone are defined by the fibre material and by the characteristics of the surface of the acid. Thus, they are influenced by the liquid above the acid. In our case silicon oil produces the best results with reference to surface smothness and reproducibility. It also forms a protective layer preventing evaporisation of the HF.

Note the hollow cone which forms because the acid corrodes the core of the fibre faster than the cladding. This is due to the doping profile of the fibre which makes the core more susceptible to the acid. Contrarily, in the second step buffered HF is used and the cladding dissolves faster. As optical fibres of different manufacturers differ in their doping profiles, both etching process and acid composition have to be finely attuned to the chosen fibre. Furthermore, not all kinds of fibres have doping profiles that are suited for our two-step etching process.

For the second step the 38 % v/v HF is buffered by 40 % v/v ammonium fluoride (NH_4F), such that one volume part each of water and HF are mixed with 7 volume parts NH_4F. Due to the different etching rates of core and cladding, the cladding around the internal cone is etched back and a small tip begins to emerge (see Fig. 3.1 (b)). This is the *secondary cone* forming the tip of the probe. Both its full opening angle (approximately 55°) and the time it takes to form are determined by the ratios between hydrofluoric acid, ammonium fluoride and water. Intermittently, the etching process is interrupted and the fibre is cleaned with n-hexane in order to monitor its progress under a microscope. It is then returned to the HF to resume the etching. After 5–9 hours the hollow cone has vanished and the process is terminated.

3.1 Probe fabrication

As hydrofluoric acid is a potently toxic contact poison, the etched fibre is subsequently cleaned thoroughly in three ultrasonic baths: first in n-hexane, then in isopropanol, and finally in double-distilled water. To prevent an accumulation of electrostatically attracted dust particles, the fibre is then placed promptly into the high vacuum coating chamber.

3.1.2 Metallisation and aperture cutting

An aluminium coating of approximately 250 nm thickness is evaporated onto the etched fibre tips in a magnetron sputtering system (BOC Edwards Auto 500). To ignite the plasma and minimise impurities, the coating chamber is evacuated to a pressure of approximately 10^{-6} mbar and cooled with liquid nitrogen. Then a flow of 5000 sccm (standard cubic centimetres per minute) of argon is passed through the chamber. For five minutes a DC gas discharge is induced with a power of 400 W, accelerating argon ions towards an aluminium target with a purity of 99.99%. There, they deliver their kinetic energy to the aluminium atoms. A magnet below the target causes the argon ions to circle around the magnetic field lines, extending the interaction length and thus enhancing the efficiency of the energy transfer to the aluminium. Then, the freed atoms condense on the fibre at a distance of approximately 9 cm from the target, producing a closed aluminium film on the probe. After the evaporation process the quality of the film is inspected visually by coupling the light of a helium-neon laser into the fibre probe. Typically, the film at the tip of the fibre reflects all the light and thus appears completely dark in this test.

While the aperture in the finished aluminium coating was formerly created by controlled squeezing against a hard glass surface, now a more reproducible and far less time-consuming technique is used: aperture-milling with a *focused ion beam* (FIB) [75]. An overview over the technique of FIB milling and further references can be found in [76]. In our group the milling is performed with a FIB unit of the Carl Zeiss AG, the "1540 EsB CrossBeam", depicted in Fig. 3.2 (a). It features a beam of ions that can be used for milling as well as for imaging. The beam consists of gallium ions and is produced by a liquid metal ion source. In addition to the gallium beam, an electron beam can be used to take images at a higher resolution of up to 2.5 nm. Importantly, this can be done simultaneously to milling so that the milling process can be monitored closely.

Fig. 3.2 (b) shows exemplary fibres on the mount for ion beam milling. With some care a dozen or more probes can be placed on it. The tips of the fibres protrude over the edge, ensuring that the probes are not damaged by any mechanical contact with the mount. However, the protrusion is kept short to restrict vibrations of the free fibre ends which are detrimental to the precision of the milling process. Patches of conductive silver connect the fibres electrically to the mount, thus preventing a build-up of the charges that are deposited during milling.

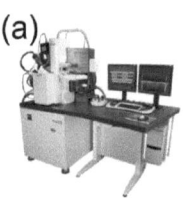 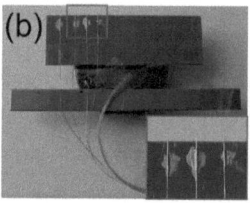 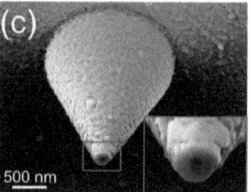

Fig. 3.2: (a) Image of a 1540 EsB CrossBeam of the Carl Zeiss AG, (b) photograph of probes on the FIB holder, and (c) typical example of a probe with a milled aperture of 140 nm.

The focused ion beam has a current of 10 pA and an ion energy of 30 keV. After careful alignment it is moved back and forth above the tip on a line perpendicular to the fibre. In order to control the size of the aperture, the line is lowered slowly towards the tip. The ion beam begins to mill away the metal at the tip of the probe, removing layer by layer until the glass is reached and an aperture forms. That aperture is allowed to grow until it has reached the desired size. Then the ion beam is stopped. A full image of the aperture is taken directly after milling, utilising a contrast enhancement due to the recently delivered charges. In a trade-off between high resolution (i.e., small apertures) and high transmittance (large ones), aperture diameters between 140 nm and 200 nm are produced (see Fig. 3.2 (c)). Note that due to the finite penetration depth of light in metal, the optical diameter is always somewhat larger than seen in the SEM. The magnitude of this effect depends on the penetration depth of the metal and on the thickness of the coating right around the aperture, which is determined by the opening angle of the probes.

The entire FIB process takes 15–20 minutes for each probe with some overhead for probe preparation, chamber evacuation, and beam alignment. In this way up to a dozen fibre probes can be processed in a single day, whereas the former process of controlled squeezing often required more than a day for a single probe.

3.1.3 Probe module

Whereas etching and aperture fabrication determine transmittance and resolution of the probe, other characteristics—such as its reliability in measurements, the ease of handling and its durability—depend on the whole unit of probe and distance sensor. This is due to the close proximity between probe and sample. It has to be controlled with high precision and reliability to achieve a stable signal and to avoid damaging the probe gradually (or, indeed, in a sudden collision).

Principle of the distance control

We use a method introduced by Karrai and Grober in [19], where a piezoelectric tuning fork is employed as distance sensor. The fibre probe is fixed on one prong of the tuning fork with its tip protruding slightly. The tuning forks used in this work are commercially utilised as a time base for quartz clocks, as they provide a particularly stable frequency of $32768\,\text{Hz} \pm 20\,\text{ppm}$. Taken from their casing they measure 6 mm by 1.5 mm by 0.34 mm. They exhibit two different vibrational modes at the same resonance frequency: in the antisymmetric mode the two prongs move in opposite directions, in the symmetric mode both move in unison. Detection schemes usually rely on the antisymmetric mode, although the forks resonate in a combination of both. They are excited mechanically with the help of a dither piezo. The amplitude of the resulting vibration has been determined to be less than a nanometre [19], hence its influence on the topographical and optical measurements is negligible. While the tuning fork resonates, the mechanical strain generates an oscillating voltage across the prongs that can be picked up *via* the prefabricated conductor paths. By using a lock-in technique, even a slight decrease in the vibration amplitude can be measured. If the protruding fibre tip comes into immediate proximity to the sample surface, such a decrease is induced by shear forces. Thus, a feed-back mechanism can be designed to keep the tip within range of those forces, which extend to about 10–20 nanometres from the sample. In our implementation of feedback control the probe is approached to the sample until its vibration amplitude drops below a preset value. It is then retracted until the amplitude exceeds the set value again. This is repeated continually, leading to a permanent vertical movement of less than five nanometres. In measurements, the noise thus induced on the optical signal is well below the noise of the detector.

Assembly

The base of the probe module is a printed circuit board with two contact pads (see Fig. 3.3, left). A small amount of glue is taken on the tip of an auxiliary fibre and placed at the rim of the board between the pads. With pincers, the edge of a tuning fork is pressed onto it and released, such that about 1 mm of the fork overlaps with the board (Fig. 3.3, centre). At this point, some experience is needed. The size of the overlap is determined by a trade-off between mechanical stability and the damping of the fork's resonance. It also depends on the type and amount of glue placed on the circuit board. The best results are achieved with an instant adhesive of the UHU GmbH & Co. KG. Furthermore, the fork should be aligned as exactly as possible perpendicularly to the board edge, so that, later, the correct angle between fibre and sample is obtained. The fork also has to be placed rapidly as the glue solidifies swiftly. When this is done, the pins of the tuning fork are soldered to the contact pads. Then, as a check on the results, the resonance curve of the fork is measured. Quality factors of at least

Fig. 3.3: Photograph of a printed circuit board prepared for assembly (left), a tuning fork glued and soldered to the board (centre), and a completed probe module with tuning fork, fibre probe, and strain relief (right).

4000 should be obtained. The forks with the highest amplitudes, lowest noise and narrowest resonance spectra are selected for further processing.

In a holder under a microscope the fibre probe is carefully aligned with the upper prong. Once more an auxiliary fibre serves to apply the glue: a tiny droplet is spread on the probe, beginning and ending at the back of the optical fibre in order to avoid dislodging it. Capillary forces then pull the probe towards the prong and draw the glue into the remaining gap. Occurring displacements have to be corrected speedily. Then the glue is left to harden for 20–30 minutes. Here again, the type of glue is selected with special care. The first requirement is a low viscosity, allowing the glue to spread into the small gap between fibre and prong. While it does so, it still has to provide a strong and close connection between probe and tuning fork, so that the shear forces acting on the probe couple to the fork's vibration with sufficient strength. At the same time the quality factor of the resonance should not drop too far. Finally, the glued connection mustn't deteriorate over time with the constant strain of the tuning fork vibration. We have found that the instant glue "Pattex Blitz Kleber flüssig" best meets all these requirements.

The fragile connection between fibre and prong is protected by a strain relief, implemented by a small piece of paper. It is firmly fixed with an epoxy resin glue ("Uhu plus sofortfest") to both the fibre and the circuit board, as well as to the pins of the tuning fork, thus providing additional stability. A completed probe module is shown on the right-hand side of Fig. 3.3.

3.2 The near-field microscope

In the following section the setup of our scanning optical near-field microscope is summarised. Detailed descriptions can be found in [56] and [77].

3.2 The near-field microscope

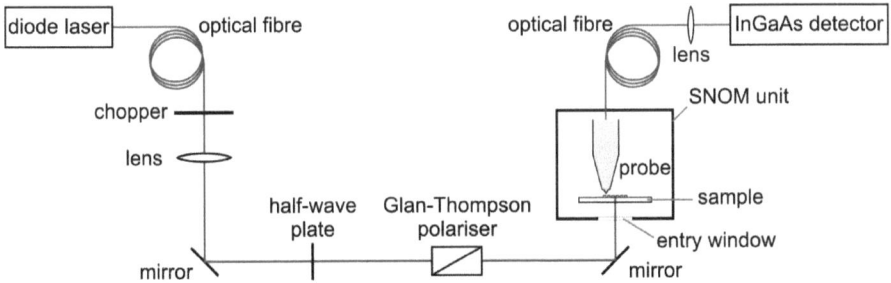

Fig. 3.4: Optical setup of the near-field microscope. Light from a diode laser is chopped for lock-in detection, focused onto the sample and polarised. Two orthogonal mirrors allow lateral positioning of the spot on the sample. The near-fields are picked up by the SNOM probe and conveyed to the detector.

3.2.1 Optical setup

The optical setup is depicted in Fig. 3.4. Six infrared diode lasers are available with wavelengths ranging from 1430 nm to 1610 nm, as well as one at 1310 nm. Additionally, for aligning purposes, a helium-neon laser at 633 nm can be employed. The light is conveyed to the setup by a single-mode optical fibre with an FC connector (SM Patchcord, Laser Components). To enable lock-in detection, it passes an optical chopper (HMS Light Beam Chopper 220) with a frequency of 1.25 kHz. It is then focused by a lens with a focal length of 50 cm and a diameter of 40 mm, which can be moved to adjust the focal plane onto the sample. The focal spot has a diameter of roughly 30 µm. Thus, within a scan window of 2 µm × 2 µm, the light intensity approximates a plane wave. Its polarisation can be turned with a combination of a half-wave plate and a Glan-Thompson polariser. Two gold mirrors at right angles allow movement in both lateral directions, so that the centre of the spot can be positioned precisely onto that area on the sample which is directly below the probe. Note, that after the reflection at the second mirror the beam travels vertically. It traverses a small entry window into the cryostat housing sample and probe. In this configuration the light passes through the substrate before it is transmitted by the sample and arrives at the probe's aperture. After this, it is guided through the probe fibre to a thermoelectrically cooled InGaAs detector with integrated preamplifier (Judson Technologies). The signal is further amplified by a lock-in detector (Model 5210, EG&G Princeton Applied Research) and recorded by the computer *via* two connector blocks (184749C-02 and BNC-2090, National Instruments).

3.2.2 SNOM unit

Fig. 3.5 shows a sketch of the SNOM unit. Although it is designed to permit the investigation of samples at low temperatures and in vacuum, these features were not used in this work. The unit is made up of the cryostat, the scanning unit and the cover.

The cryostat is the bottom part of the SNOM. It includes the aforementioned entry window, the sample holder on a heat exchanger, a shield against thermal radiation featuring an access hole for the probe, and a duct for liquid helium cooling. The heat exchanger and sample holder can comfortably be positioned laterally by a micro-positioning system (SMC basic, miCos GmbH), which moves the sample while maintaining the probe and the laser spot fixed, thus permitting an examination of different parts of the sample without readjustment of the focus position. A small mirror next to the sample allows a coarse visual inspection of the sample and of the probe's location with respect to it. For this purpose a stereomicroscope (Leica MZ8) is located above the SNOM.

The scanning unit serves as the top of the cryostat. Its central element is the 3-D scanning piezo (Tritor 101, piezosystem jena GmbH), which is set on a hinge above the base plate, and is supported by two springs, so that the whole piezo–probe assembly can be pushed towards the sample with the help of a micrometre screw. This coarse approach can be monitored with the stereomicroscope and is utilised to bring the probe to within roughly 20 µm of the sample. The high-precision part of the approach is carried out by the 3-D piezo. It covers a large scanning range of 100 µm in all three directions. Its movement is directed by a specifically tailored LabView program that carries out both the raster scan across the sample and the distance feedback control. On the piezo's lower side the probe module (see also right-hand side of Fig. 3.3) is located. If necessary, it can be adjusted laterally with an $r\phi$ coarse-positioning system. It is held by copper clips which also provide the electrical contact to the tuning fork electrodes. The signal from these electrodes is amplified by a lock-in amplifier (Model SR830 DSP, Stanford Research Systems) and directed to the computer as feedback for the distance control. The small dither piezo used for the excitation of the tuning fork resonance sits directly behind the probe module's circuit board. It is actuated by a frequency generator (33120A, Hewlett-Packard). The small separation distance between dither piezo and tuning fork ensures that the fork is excited strongly and reliably. The probe fibre, as well as the wiring for both the dither piezo and the tuning fork amplitude detection pass through the central hole of the scanning piezo.

The cover of the SNOM encloses the scanning unit. It accommodates the micrometre screw for the coarse approach, a passage for the probe fibre, contacts for the wiring of the piezos (scanning piezo, dither piezo and piezoelectric tuning fork), and the window for visual inspection by the

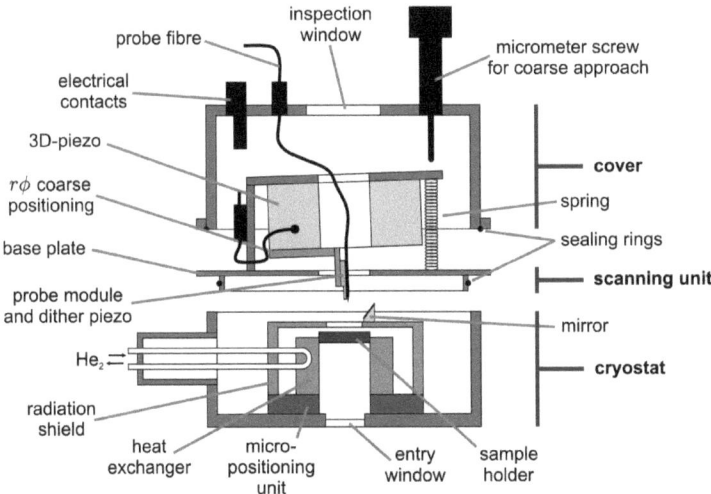

Fig. 3.5: Sketch of the SNOM unit: cryostat, scanning unit, and cover. Adapted from [78].

stereomicroscope. This window is blocked during measurements to prevent stray light from reaching the detector.

3.3 Sample fabrication

Several different metamaterial samples were used in this work: double-wires (fabricated by Stefan Linden), a fishnet structure [79] (fabricated by Gunnar Dolling) and split-ring-resonator arrays [80] (by Manuel Decker). Their design is an iterative process of fabrication, characterisation and modification of the desired metamaterial. In the following sections the fabrication is detailed, based on the description in [41]. An overview over the steps employed is given in Fig. 3.6.

Substrate preparation

For the fabrication of the metamaterial samples we use fused-silica ("Suprasil") substrates measuring 10 mm × 10 mm × 1 mm. First of all they are cleaned thoroughly with acetone and with isopropanol. They are then covered with 5 nm of the transparent conductor *indium-tin-oxide* (ITO). This is done in an electron-beam evaporation chamber at $1.2 \cdot 10^{-5}$ mbar. The ITO layer later prevents the electrostatic build-up of charges on the sample and promotes the adhesion of the gold film, while at the same time being suitably transparent at optical

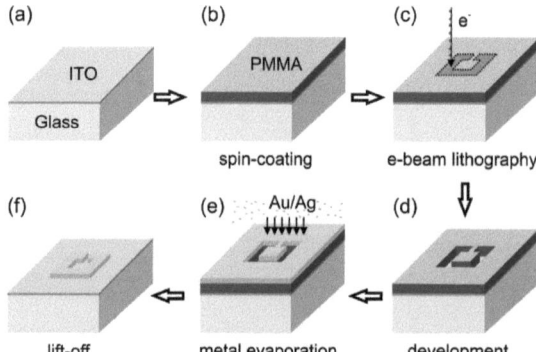

Fig. 3.6: Fabrication of metamaterial samples, taken from [81]. A glass substrate is covered with ITO (a) and spin-coated with the photoresist PMMA (b). Using electron-beam lithography, 2-D structures are "written" into the resist (c) and removed by a developer (d). Then, gold is evaporated onto the sample and onto the PMMA-free areas (e). Finally, the remaining PMMA is lifted off the sample, along with the gold film covering it (f), leaving gold only at the formerly exposed parts.

frequencies. For 5 hours, the covered substrate is post-baked at a temperature of 450 °C. Subsequently, it is spin-coated for 90 seconds at 5000 rpm with a 4% solution of *poly(methyl methacrylate)* (PMMA, MicroChem Corp.)—a commonly used positive photoresist—in anisole. This results in a PMMA film of approximately 200 nm thickness, which is post-baked for 30 minutes at a temperature of 165 °C.

Electron-beam lithography

The PMMA resist is made up of long polymer chains that can be broken apart by high-energy electrons if the charge dosage exceeds 175 µC/cm². This is done in a *scanning electron microscope* (SEM), where an electron beam is directed onto the resist. It is moved by a beam-deflection unit in conjunction with a high-speed pattern generator and a beam blanker controlled by a CAD software. With these, almost arbitrary two-dimensional patterns can be "written" into the photoresist.

Two electron microscopes were employed to fabricate the samples studied in this work: a Zeiss "Supra 55 VP" SEM with an external pattern generation hardware ("ELPHY Plus", Raith GmbH) for the fishnet structure, as well as the ultra-high-resolution electron-beam-lithography system "e-LiNE" by Raith GmbH for the double-wire and SRR samples. Electron energies of 10–30 keV were used.

The best writing resolution attainable with these systems is lower than that of SEM images taken on the same machines. This is mainly due to the secondary electrons that are generated

3.3 Sample fabrication

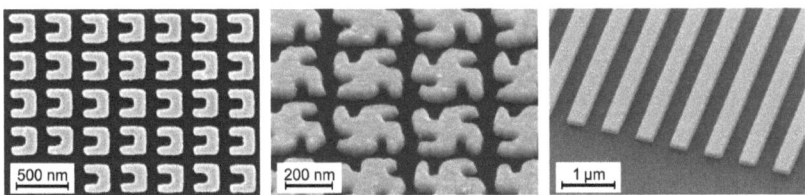

Fig. 3.7: Examples of metamaterials fabricated with electron-beam lithography: (a) a split-ring-resonator array, (b) right-handed gammadions, and (c) the edge of a double-wire structure. (a) and (b) courtesy of Manuel Decker.

where the electron beam meets the photoresist. They enlarge the effective spot size and, moreover, increase the dosage in the area around the beam. Thus, in conjunction with secondary electrons of features written nearby, the threshold dosage for exposing the resist can be exceeded outside the written structures. In addition to this so-called *proximity effect*, aberrations have an influence on the smallest obtainable feature size: unless they are corrected by a specific lens system, astigmatisms cause an elliptic spot whenever the beam deviates significantly from the optical axis. This leads to a loss of resolution both in the images taken with the SEM and in the written structures.

With careful adjustment of the electron-beam dosage to reduce the impact of the proximity effect, and with proper lens corrections for astigmatism, feature sizes as small as 10–20 nm can be created with electron-beam lithography.

Post-processing

After lithography the sample is submerged for 20 seconds into a developer, a 3:1 mixture of isopropanol and methyl isobutyl ketone (MIBK). This mixture dissolves all areas where the resist has been exposed, leaving a negative mask of PMMA. Then, the sample is processed in the electron-beam evaporation chamber. At pressures below 10^{-6} mbar a layer of gold is evaporated onto the mask. If required, further layers of various materials are added. Finally, the sample is treated with hot acetone in an ultrasonic bath, where the remaining PMMA is removed. In this *lift-off*, the parts of the gold and the additional layers which lie above the PMMA are removed as well. Where the resist was cured by the electron-beam lithography, however, gold remains on the substrate, forming the finished metamaterial structure. Examples of metamaterials produced in this fashion are shown in Fig. 3.7.

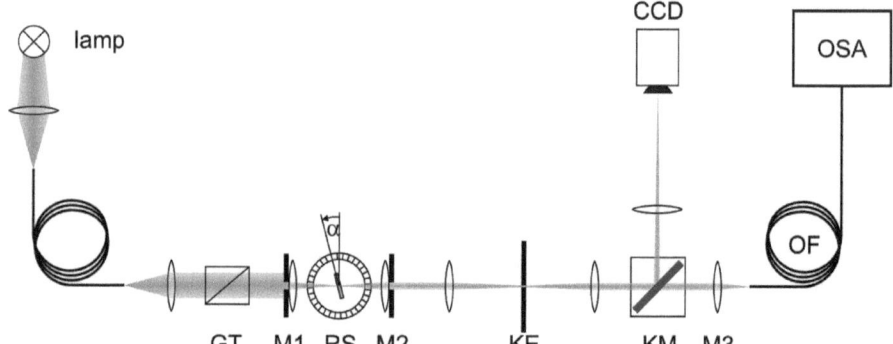

Fig. 3.8: Sketch of the setup used for spectrally resolved, far-field transmittance measurements. GT: Glan-Thompson polariser; M 1–3: microscope objectives; KA: knife edge; KM: kinematic mirror; OF: optical fibre; and OSA: optical spectrum analyser. Adapted from [49].

3.4 Transmittance characterisation

In order to characterise the fabricated samples, spectrally resolved measurements of their far-field transmittance were conducted in this work. A home-built setup from our group was used [82], which is described in great detail in [49]. It is shown in Fig. 3.8. In this setup the light of a 100 W tungsten halogen lamp is collimated and polarised with a Glan-Thompson polariser ("GT", Melles-Griot 03PTO001). It is then focused onto the sample by a microscope objective with a circular aperture ("M1"). Coming from the substrate side it passes through the sample which is set on a rotation stage ("RS"), and is focused again by a second microscope objective, "M2", onto an intermediate focal plane. Here, a knife edge aperture ("KE") is used to restrict the measured area. Behind this, a kinematic mirror ("KM") can be placed into the light path to divert the beam onto a CCD camera, where an image of the sample is created. This can be used to comfortably adjust the focus onto the sample and select the desired positions for the transmittance measurements. To detect the light, the kinematic mirror is removed and the beam is coupled into an optical fibre ("OF") that leads to the *optical spectrum analyser* ("OSA", Ando AQ-6315B, 350~1750 nm).

For the samples examined in this work we conducted the far-field characterisation with the OSA setup under normal incidence ($\alpha = 0$ in Fig. 3.8). All resulting transmittance spectra were referenced with respect to transmittance spectra of the bare sample.

3.5 Phase-sensitive near-field microscope

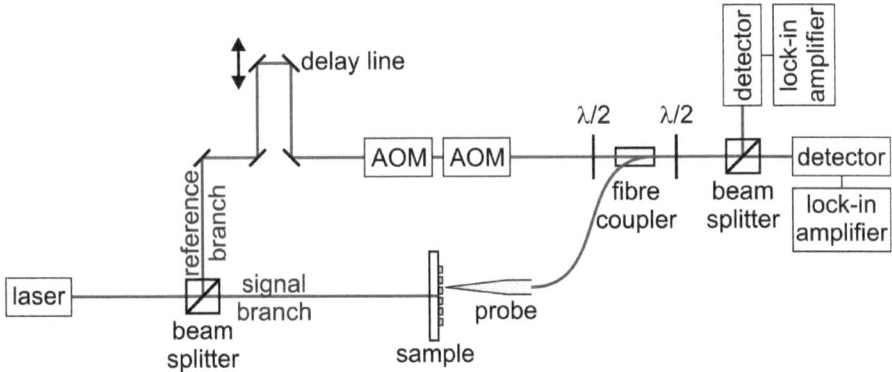

Fig. 3.9: Interferometric, heterodyne, phase-sensitive near-field microscope of the group of Prof. Kuipers. The reference beam is frequency-shifted before it interferes with the signal from the near-field probe. A polarising beam splitter separates the orthogonal polarisation directions and sends them to two different detectors. Based on [84].

3.5 Phase-sensitive near-field microscope

The measurements discussed in chapters 5 and 6 were conducted in cooperation with the group of Prof. Kuipers of the FOM Institute AMOLF (Foundation for Fundamental Research on Matter Institute for Atomic and Molecular Physics) in Amsterdam. They were performed by Matteo Burresi with the phase- and polarisation-sensitive, heterodyne near-field microscope of that group. A detailed description and analysis of the setup without the polarisation separation is given in [83]. The overview given here loosely follows [84].

Fig. 3.9 sketches the setup of the phase-sensitive near-field microscope as used in the measurements. The laser beam is split into a signal and a reference arm. In the signal arm, the light illuminates the sample. There, the resulting near-fields are picked up by the fibre probe. If handled properly, the probe fibre maintains the orthogonal polarisation states of the signal. It conveys them to a fibre coupler, where they interfere with the light from the reference arm.

The reference beam passes a delay line, with which the length of the reference branch can be adjusted to match the length of the signal branch. In a heterodyne scheme the reference beam frequency is doppler-shifted by two *acousto-optic modulators* (AOMs). One induces a shift of 80 MHz, the other an opposite shift of 80.04 MHz, resulting in a total shift of $\Delta\omega = \pm 2\pi \cdot 40$ kHz. These opposite shifts are necessary because the minimum shift induced by a single AOM exceeds 40 kHz by several orders of magnitude. Subsequently, before the beam enters the fibre coupler, its polarisation is rotated with a half-wave plate, so that it interferes equally with both polarisation states of the signal branch.

With the electric fields of the reference and signal branches denoted by $\boldsymbol{E}_R$ and $\boldsymbol{E}_S$, respectively, the intensity after the coupler can be calculated as

$$I \propto (\boldsymbol{E}_R + \boldsymbol{E}_S) \cdot (\boldsymbol{E}_R + \boldsymbol{E}_S)^* \qquad (3.1a)$$
$$\propto |\boldsymbol{E}_R|^2 + |\boldsymbol{E}_S|^2 + 2\boldsymbol{E}_R \cdot \boldsymbol{E}_S \cos(\Delta\omega t + \Delta\phi), \qquad (3.1b)$$

where $\Delta\phi$ is the phase difference between signal and reference beam. The last term of Eq. (3.1b) oscillates with a frequency of $\Delta\omega$. This is used for noise reduction by lock-in amplification. The amplitude registered by the lock-in scheme is the amplitude of the last term, that is, $2\boldsymbol{E}_R \cdot \boldsymbol{E}_S$. Thus, the detected signal is proportional to the electric field of the beam from the signal branch, $\boldsymbol{E}_S$. Importantly, the phase difference between reference and signal branch, $\Delta\phi$, is retrieved as well in the lock-in scheme, allowing the experimental determination of the phase acquired on transmission through the sample. Another advantage of the detection scheme can be shown by introducing the *heterodyne gain*, $\gamma = |\boldsymbol{E}_R|/|\boldsymbol{E}_S|$. Then, $2\boldsymbol{E}_R \cdot \boldsymbol{E}_S = 2\gamma |\boldsymbol{E}_S|^2$ and it becomes apparent that the signal-to-noise ratio can be increased considerably if $|\boldsymbol{E}_R| \gg |\boldsymbol{E}_S|$ is chosen.

After the interferometric signal leaves the fibre coupler and before it reaches the detectors, its polarisation is rotated by a second half-wave plate and it passes through a polarising beam splitter. Half-wave plate and beam splitter are oriented such that the polarisation states of the signal are separated and sent to two different detectors and lock-in amplifiers. Thus, both polarisation states of the signal from the near-field probe are detected separately, and can be analysed individually.

Prior to measurements, the length of the delay line and the phase difference $\Delta\phi$ are adjusted with respect to the bare substrate. The two polarisation directions are chosen parallel and orthogonal to the incident laser polarisation. According to Sandtke *et al.* [83], the registered phase difference is subject to a drift of approximately 0.3 degrees per second. Measurements should, therefore, be conducted speedily.

The probes used in the group of Prof. Kuipers are manufactured by heating and pulling. Therefore, the produced opening angles are merely 18 degrees wide (full opening angle), leading to a rather low transmittance. However, the losses thus incurred are more than compensated for by the high gain due to the heterodyne detection scheme.

4 Model for the near-field probe

In principle, three-dimensional field calculations can provide the transmission through a near-field probe positioned over a particular point of the sample. However, in order to obtain a two-dimensional image of the sample, one such 3-D calculation would have to be executed for every single pixel. As each individual calculation is extremely time-consuming on its own, this method is hardly feasible for our investigations. Other approaches involve restriction to two dimensions and/or complex models and sophisticated analytical calculations to account for the interaction between probe and sample (see for example [23, 24, 85–88] and references therein).

A possible simplification is to disregard multiple scattering between probe and sample [25]. In this approach the probe is assumed to detect the near-field on the surface of the metamaterial without disturbing it in any way and a *transfer function* for the near-field probe is derived ([26], for scanning tunnelling optical microscopes). The strength of the probe–sample interaction is difficult to determine beforehand as it depends on the structure under investigation and a number of experimental details, *e.g.* the size and geometry of the probe, its distance to the sample, and the wavelength used. Caution, therefore, has to be exercised in comparing the results of such calculations to the corresponding measurements. However, in all cases studied in this work we find a strong agreement between the calculated and measured images, indicating that the interaction effects are indeed weak enough to be neglected. In the following sections we present our model for the imaging process of an aperture-probe near-field microscope (chapter 4.1), discuss its implementation (chapter 4.2), and show some exemplary results (chapter 4.3).

4.1 Theoretical description of the imaging process

Disregarding the influence the probe exerts on the sample allows for a much faster and simpler method to numerically derive two-dimensional SNOM images [79, 80]: we consider the near-fields on the sample as an uninfluenced *source* and regard the SNOM probe as a *coupler* which links the source fields to the guided fields in the probe fibre, the *waveguide*. Thus, in a first step the near-field probe is ignored and a mostly straightforward computation of the electromagnetic fields in the metamaterial sample is executed. With this, the vectorial field distributions of E and H are obtained. In the second step we numerically simulate the action of the coupler.

4 Model for the near-field probe

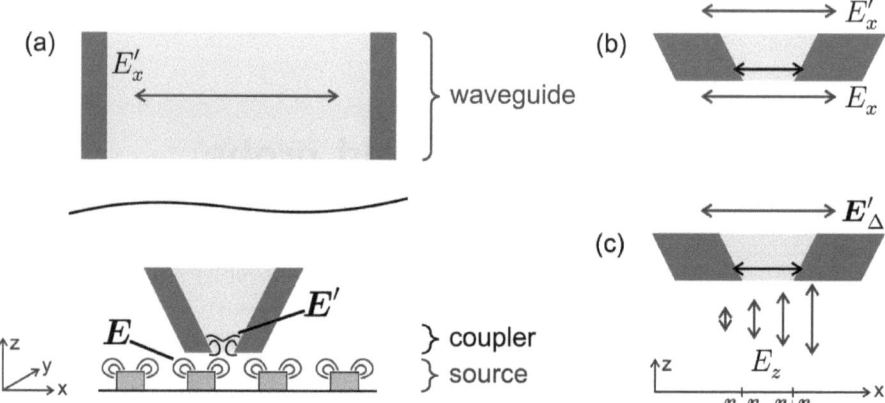

Fig. 4.1: (a) The near-fields (E, sketched in blue) are transmitted by the aperture (E', sketched in black) and couple to guided modes in the probe fibre (sketched in green). Coupling of electric dipole oscillations (b), and of the spatial gradient of E_z (c).

In order to do this, a model has to be found that describes (i) the transition of near-fields through the aperture and (ii) the conversion of the transmitted fields into waves guided along the probe fibre towards the detector. The difficulty here lies, firstly, in the selection of the decisive components of the complex electric and magnetic fields and, secondly, in the correct way to combine them. The following reasoning, parts of which were developed in collaboration with the group of Prof. Kuipers of the FOM Institute AMOLF (Foundation for Fundamental Research on Matter Institute for Atomic and Molecular Physics) in Amsterdam, motivates the choices made in this work.

Let us first consider the conversion of the transmitted electric fields into guided waves. As illustrated in Fig. 4.1 (a), the electric field of the guided waves (sketched in green for the x-direction) is oriented perpendicularly to the direction of propagation, the z-direction. This means, that of the fields transmitted through the aperture, E' (sketched in black), only the x- and y-components will couple into the waveguide and reach the detector. These transversal electric fields are generated by correspondingly oriented electric dipole oscillations, *i.e.*, oscillating build-ups of charges across the aperture. Note that the magnetic fields need not be taken into account separately, because, as described by Maxwell's equations, they generate—and are in turn generated by—the electric fields. Thus, their influence is already comprised in the electric fields. By considering the magnetic fields the same results are obtained as by considering the electric fields. We have verified this for all the calculated field distributions shown in this work (see below).

4.1 Theoretical description of the imaging process

For the transition through the aperture, namely from $\boldsymbol{E}$ to $\boldsymbol{E}'$ (see Fig. 4.1 (a)), we therefore only have to include the electric dipole field. It can be induced by the transversal components of the near-field, E_x (Fig. 4.1 (b)) and E_y. Due to the finite width of the aperture they are averaged over its extension, the area A. Importantly, an oscillating electric field, $\boldsymbol{E}'_\Delta$, between the opposite sides of the aperture's rim can also be caused by a spatial gradient of E_z (shown in Fig. 4.1 (c) for the x-direction). For each pair of opposing points around the aperture, the strength of this "gradient field" is proportional to the difference of E_z between them. Its direction is parallel to the vector connecting the points, $2\boldsymbol{r}_\alpha$, where α is the angle between this vector and the positive x-axis. Thus, with the location of the aperture's centre denoted by $\boldsymbol{r}$, this yields:

$$\boldsymbol{E}'_\Delta(\alpha) \propto (E_z(\boldsymbol{r}+\boldsymbol{r}_\alpha) - E_z(\boldsymbol{r}-\boldsymbol{r}_\alpha)) \cdot \hat{\boldsymbol{e}}_{r_\alpha}. \qquad (4.1)$$

This should not be confused with the local partial derivatives, $\partial E_z/\partial x$ and $\partial E_z/\partial y$, representing the gradients at a single point in space. To obtain the total contribution from the near-field component E_z to the components E'_x and E'_y on the other side of the aperture, $\boldsymbol{E}'_\Delta$ is projected onto the respective direction (x or y) and averaged over the entire rim. We will designate this integrated contribution by $\Delta_x E_z$ and $\Delta_y E_z$ for the x- and y-direction, respectively:

$$\Delta_x E_z = \int_{-\frac{\pi}{2}}^{+\frac{\pi}{2}} \boldsymbol{E}'_\Delta(\alpha) \cdot \hat{\boldsymbol{e}}_x \, d\alpha, \qquad (4.2a)$$

$$\Delta_y E_z = \int_{-\frac{\pi}{2}}^{+\frac{\pi}{2}} \boldsymbol{E}'_\Delta(\alpha) \cdot \hat{\boldsymbol{e}}_y \, d\alpha. \qquad (4.2b)$$

Note, that the gradient contributions are not necessarily transmitted with the same phase and amplitude as E_x and E_y. This is accounted for by multiplication with a complex coefficient C. Thus, we obtain for the transmitted fields:

$$E'_x \propto \frac{1}{A}\int_A E_x \, dA + C\Delta_x E_z \quad \text{and} \qquad (4.3a)$$

$$E'_y \propto \frac{1}{A}\int_A E_y \, dA + C\Delta_y E_z. \qquad (4.3b)$$

This is done for every position $\boldsymbol{r}$ of the probe. The normalised intensity distribution then reads

$$\frac{I(\boldsymbol{r})}{I_{\text{ref}}} = \frac{1}{C_{\text{ref}}}\left(|E'_x(\boldsymbol{r})|^2 + |E'_y(\boldsymbol{r})|^2\right), \qquad (4.4)$$

where C_{ref} is a normalisation constant determined with a reference calculation (see section 4.2.2).

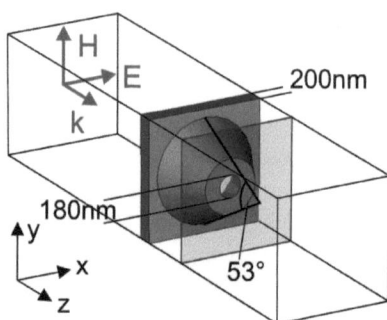

Fig. 4.2: Sketched cross section of the model used to calculate the transmission characteristics of our near-field probes.

Numerical corroboration

In order to substantiate our reasoning, we use Lorentz' reciprocity theorem: we turn around the propagation direction and investigate numerically how the guided waves behind the aperture couple to the near-fields in front of it. To this end, calculations were executed with the numerical simulation tool Microwave Studio (introduced in chapter 2.3.1). The cross section of the configuration is sketched in Fig. 4.2. We computed the transmission of a plane wave with a wavelength of 1510 nm, polarised in x-direction, through an aluminium-coated glass cone with $n_{glass} = 1.5$, a total opening angle of 53° and an aperture diameter of 180 nm. For the 200 nm aluminium coating the Drude model was used, with parameters taken from [89], namely $\omega_{pl} = 2.24 \cdot 10^{16}\,\text{s}^{-1}$ and $\omega_c = 1.22 \cdot 10^{14}\,\text{s}^{-1}$. Boundary conditions were "open" (i.e., perfectly matched layers) in the propagation direction z, and perfect electric and magnetic conductor boundaries for x and y, respectively. We find that the overall transmittance of the model, $P_{out}/P_{in} = 2.4 \cdot 10^{-6}$, reproduces a typical experimental value for the transmittance of near-field probes. As indicated in Fig. 4.2 (transparent blue plane), the generated near-fields were analysed on a plane at 20 nm distance from the probe aperture.

The amplitudes of the resulting field distributions are shown in Fig. 4.3 (a), the corresponding phase distributions are depicted in Fig. 4.3 (b). The dashed circles indicate the position and size of the aperture in both the phase and the amplitude images. Obviously, the agreement with our earlier reasoning is excellent: as expected, the impinging plane wave couples strongly to the x-component of the electric field behind the aperture, as well as to the z-component. Here, opposite points on the aperture's rim exhibit a phase difference of π, resulting in the "gradient contribution" discussed above. Furthermore, and again as expected, the coupling of the plane wave polarised in x-direction to E_y is weaker by more than an order of magnitude

4.1 Theoretical description of the imaging process

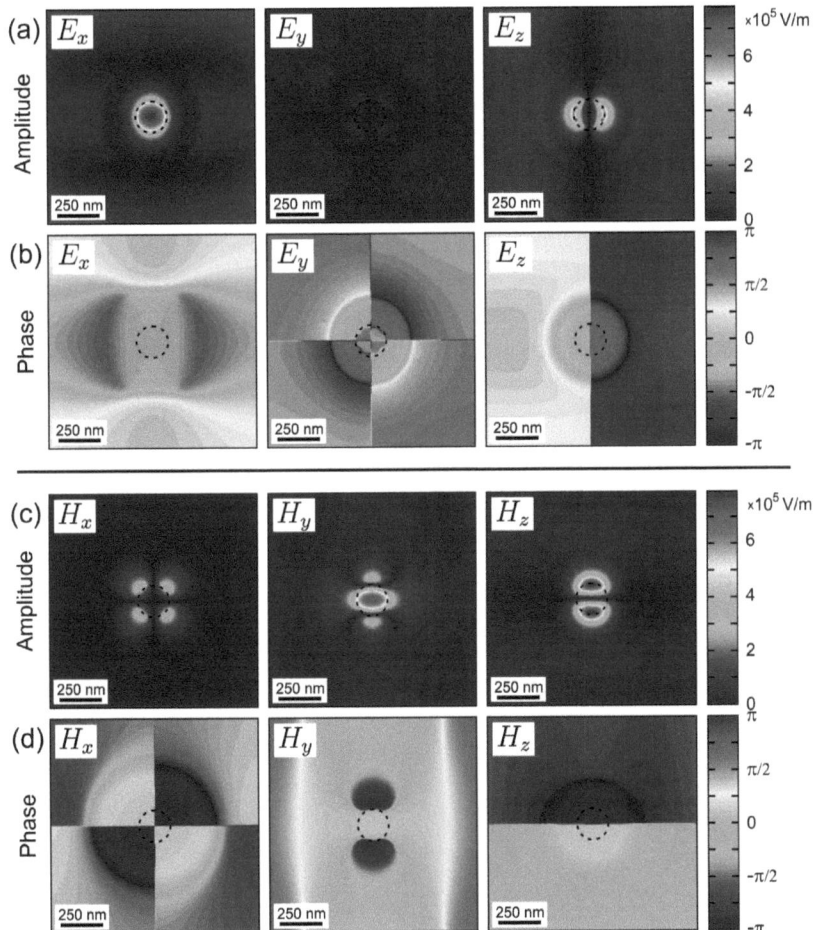

Fig. 4.3: Amplitude and phase of calculated field distributions 20 nm behind the aperture of the probe. The aperture is indicated by the dashed circles, the model employed is sketched in Fig. 4.2.

than the coupling to E_x and E_z. Indeed, consideration of the system's symmetry leads us to conclude that any non-zero E_y is a numerical artefact.

Additionally, we repeated the same calculations at different wavelengths in a range from 1450 nm to 1650 nm (not shown). The resulting field distributions are nearly identical, differing only slightly with respect to the extension of the spots around the aperture. However, a close inspection also reveals a small change in the phase difference between E_x and the gradient contribution from E_z, which varies from roughly -0.2π at short wavelengths to $\approx 0.25\pi$ at longer wavelengths. As noted above, this phase is accounted for in Eqs. 4.3 by the complex coefficient C. Thus, we expect C to depend on the wavelength.

For the sake of completeness the magnetic fields at 1510 nm are shown, too, in Figs. 4.3 (c) and (d). If the distributions of H_x, H_y and H_z are used for the numerical probe simulations, the results obtained are almost identical to those produced in the usual way, *i.e.*, with the electric field distributions. As already mentioned above, we have verified this for all the calculated field distributions shown in this work. Exemplarily, we compare two magnetic-field derived images to electric-field derived images in section 4.3, Fig. 4.5. Due to the strong similarity, in most cases treated in this thesis no additional insights would be gained from displaying the magnetic-field derived signal. Thus, we then refrain from presenting them. The only exception to this is discussed in chapter 5.

4.2 Numerical implementation

Two commercially available software packages were used to implement the model described above. The computation of the electromagnetic fields in the measured structures was performed with CST Microwave Studio. The software package "Matlab" (by The MathWorks Inc.) was employed to realise the home-made algorithms. These include the scattering-matrix code for the field distribution of the single-wire structure (implemented by Stefan Linden) and the code computing the transmission through the probe (described below). The following sections specify the calculation details.

4.2.1 Field distribution calculations

Microwave Studio

We calculated the three-dimensional vectorial field distribution of $\boldsymbol{E}$ and $\boldsymbol{H}$ in all measured samples with Microwave Studio, which uses the finite-integration technique introduced in chapter 2.3.1. The computational domain, a single unit cell, was discretised by a hexahedral grid. Its boundaries were periodic in both lateral directions (x and y) and "open"—that is, *perfectly*

4.2 Numerical implementation

matched layers (PMLs)—in the propagation direction (z and $-z$). We used a refractive index of $n = 1.45$ for the substrate and $n = 1.38$ for the MgF_2 in the double-wire and fishnet structures. The material parameters of the gold were defined through the Drude model with a plasma frequency and collision frequency of $\omega_{pl} = 1.37 \cdot 10^{16}\,s^{-1}$ and $\omega_c = 1.4 \cdot 10^{14}\,s^{-1}$, respectively. The value for the collision frequency deviates slightly from that used in chapter 2.1.4. It was chosen to represent more closely the characteristics of films of gold that are only tens of nanometres thick. The metamaterials were illuminated through the substrate by a "plane wave" pulse with frequencies from 0 to 300 THz in a Gaussian envelope (for the double-wires we used 0–250 THz). The spatial sampling rate was 5% of the shortest wavelength used and the calculation was terminated as soon as the energy inside the calculation domain had fallen below -60 dB of the maximum energy.

The resulting field distributions comprise six complex values (one each for the three components of both $\boldsymbol{E}$ and $\boldsymbol{H}$) at every discretisation point in the computational domain and for every preset frequency. However, only the near-field data at the approximate scanning height of the probe was exported for the second part of the implementation. For quality control, far-field transmittance and reflectance spectra were also recorded and compared to the measured spectra. Finally, for each sample an additional calculation was performed as a reference for the transmittance of the probe, encompassing only the glass substrate without any metamaterial structure on top of it.

Scattering-matrix

To test our model against earlier measurements [46, 78, 90], we calculated the SNOM signal expected from single-wire structures (see section 4.3). In these structures, waveguiding effects occur, caused by an ITO layer between substrate and wires. Microwave Studio fails to reproduce these effects, so a scattering-matrix code implemented by Stefan Linden (see chapter 2.3.2) was employed for the first calculation step. The second calculation step implementing the probe model was executed in the same manner as all other numerical probe simulations presented in this thesis (see below). For the single-wire structure, too, we performed a reference calculation of a bare substrate.

4.2.2 Transmission through the probe

The probe model was implemented in a home-made Matlab code [79, 80]. It uses the two-dimensional near-field data obtained before (with Microwave Studio or with the scattering-matrix code) to simulate a SNOM image of the metamaterial. For every pixel of this image—that is, for each position of the probe over the sample—the fields transmitted through the

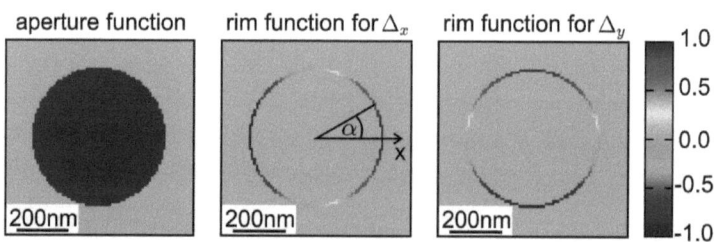

Fig. 4.4: (a) Aperture function used for the transmission of E_x and E_y through the probe. (b) and (c) Rim functions used to compute $\Delta_x E_z$ and $\Delta_y E_z$, respectively. The aperture diameter is 280 nm.

probe are derived from the near-field data through Eqs. (4.3). In the following paragraph we will consider only the calculation of E'_x, as that of E'_y is obtained analogously.

The contribution from E_x to the transmitted field E'_x is the integral in Eq. (4.3a). It is computed by multiplying E_x with an aperture function that equals one inside the aperture and zero everywhere else (shown in Fig. 4.4 (a)). Throughout this work we used an aperture diameter of 280 nm. Following Eq. (4.3a), the product of E_x and the aperture function is then integrated and divided by the area of the aperture. The contribution from E_z is calculated similarly: firstly, the *rim function* is generated (see Fig. 4.4 (b) and (c)). It is zero everywhere except at the rim of the probe's aperture. Its values at the rim run from -1 to 1 according to $\cos \alpha$ ($\sin \alpha$ for Δ_y), where α is the angle between the positive x-axis and the line connecting the aperture's centre to the point on the rim. Secondly, E_z is multiplied with the rim function. This way, both the subtraction in Eq. (4.1) and the projection onto the x-direction are executed. Thirdly, the average is calculated as with E_x: the product of E_z and the rim function is integrated and divided by the area of the rim. Note here, that the area of the rim is non-zero due to the discretisation.

Now, for every pixel of the calculated two-dimensional SNOM image the contributions from E_x, E_y and E_z are combined according to Eqs. (4.3) to obtain $E'_x(\boldsymbol{r})$ and $E'_y(\boldsymbol{r})$. The complex factor C in these equations was determined phenomenologically, that is, by optimisation of the agreement between experiment and theory. For the probes in our group it is $0.5 \cdot \mathrm{e}^{\mathrm{i} \cdot 0.3 \pi}$ at the wavelength used in our measurements (1510 nm). For the quite differently shaped probes of the group of Prof. Kuipers we found $C = 2.7 \cdot \mathrm{e}^{-\mathrm{i} \cdot 0.12 \pi}$ at a wavelength of 1510 nm. In the same way and with the same parameter C as before, the reference calculations of a bare substrate are processed to give $E'_{x,\mathrm{ref}}(\boldsymbol{r})$ and $E'_{y,\mathrm{ref}}(\boldsymbol{r})$. These are averaged over a unit cell and yield the reference constant $C_{\mathrm{ref}} = |\overline{E'_{x,\mathrm{ref}}}|^2 + |\overline{E'_{y,\mathrm{ref}}}|^2$. Finally, by adding the modulus squared of $E'_x(\boldsymbol{r})$ and $E'_y(\boldsymbol{r})$ and subsequently dividing by the reference constant C_{ref}, the normalised intensity

I/I_{ref} at the detector can be calculated (see Eq. 4.4). Locally, this referenced intensity signal can attain values greater than one.

The near-field microscope of Prof. Kuipers' group is polarisation- and phase-sensitive and, as described in chapter 3.5, measures the complex electric field picked up by the probe. Thus, results from measurements performed with it have to be compared directly to the amplitude and phase of the referenced transmitted fields, $E'_x/\overline{E'_{x,\text{ref}}}$ and $E'_y/\overline{E'_{y,\text{ref}}}$. Note that the references here are complex and, thus, include a reference phase. Similarly to the referenced intensity signal the referenced amplitude signal can exceed unity.

Obviously, the aperture function and rim functions used here to simulate the action of the probe are rather simple approximations. A comparison of Figs. 4.4 and 4.3 suggests that in order to represent the gradient contributions correctly, the model has to be implemented with a larger aperture diameter than was used for the corresponding measurements. In addition, the aperture function neglects the variation of intensity across the aperture. Furthermore, as mentioned before, in the transmission calculation the spatial extension of the spots and their phase difference depend on the wavelength used. Thus, the complex factor C balancing the strength and phase of the different contributions has to be chosen with care. We found that the general appearance of the simulated signal starts to change significantly if the amplitude or phase of C, respectively, are multiplied or divided by a factor of roughly 2, or 1.2.

Ideally—if they were known—frequency-dependent Green's functions for the coupling should be used in the probe model. As they inherently include the correct phase and amplitude relations, the coupling factor C would then be unnecessary.

4.3 Exemplary results

To test our probe model we compared results obtained with it to measurements on single-wire structures (introduced in chapter 2.2.2). The measurements were conducted earlier in our group by Nicole Neuberth and Uli Neuberth [46, 78, 90]. They are shown in the left-hand side column of Fig. 4.5, with the near-field intensity signal across the wires plotted against wavelength. The single-wire structure is sketched in the inset. In (a) the width of the wires is 115 nm and the ITO film is 30 nm thick. In (b) the wires are 80 nm wide and the thickness of the ITO layer is 140 nm. In both samples the wires are 20 nm high and repeat with a lattice constant of 425 nm. These dimensions were used for the S-matrix calculations. Furthermore, we utilised experimental data from [31] for the permittivity of gold (shown in Fig. 2.1), $n = 1.38$ for the MgF_2 and permittivities of $n = 1.46$ and 1.9, respectively, for the glass substrate and the layer of ITO on top of it. The plane-wave expansion was truncated after 301 expansion orders and we evaluated wavelengths from 480 nm to 830 nm. As the measurements were carried

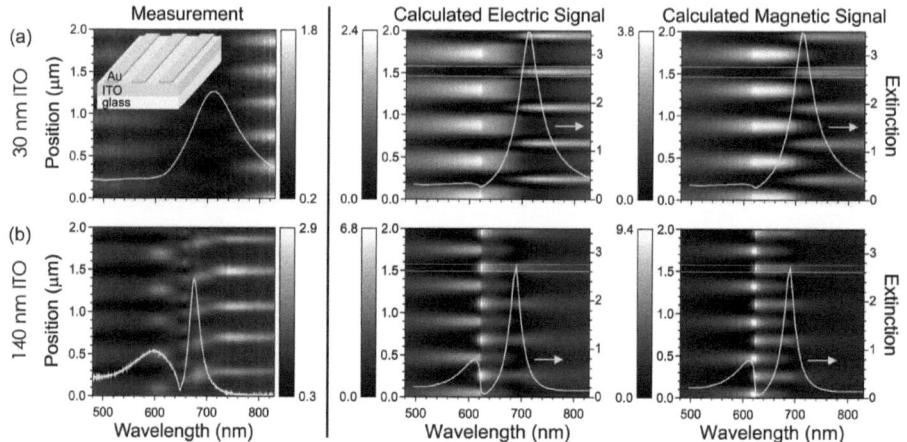

Fig. 4.5: Left-hand side column: measured extinction spectra (yellow lines, arbitrary units) and near-field distributions of two single-wire structures on (a) a 30 nm and (b) a 140 nm thick ITO waveguide. Taken from [90]. The inset shows a sketch of the structure (not to scale). Center and right-hand side column: corresponding calculated extinction spectra and near-field distributions. The position of one wire is indicated in white.

out at visible wavelengths, much smaller probe apertures were used than in the near-infrared measurements conducted for this thesis. Thus, we implemented the probe model with a different complex factor $C = 1.6 \cdot e^{i \cdot 0.35 \pi}$, and an aperture diameter of only 160 nm. Resulting near-field distributions are displayed in the center column of Fig. 4.5, with the position of one wire marked in white.

In the investigated range of wavelengths the 30 nm thick layer of ITO supports no guided mode. The corresponding spectrum, therefore, only shows the resonance of the wires at a wavelength of approximately 700 nm. The corresponding near-field distribution exhibits a transition from dark areas on top of the wires and bright areas between them at short wavelengths to a reversed distribution at long wavelengths. This contrast inversion is replicated in the calculated image.

Of greater interest is the structure with the 140 nm thick ITO layer, which acts as a waveguide in this frequency range. As mentioned on page 15, the Bragg resonance of the waveguide and the fundamental resonance of the single wires show an avoided crossing. Consequently, in the extinction spectrum the two resonances are separated by a sharp dip. When investigated in the near-field, the field distribution likewise exhibits a strong dependence on frequency both in the experiment and in the calculation. The contrast reversal seen in the 30 nm structure can be observed here as well, and is also reproduced in the calculated data. In addition, at the wavelength of the sharp extinction dip two rather than one bright spots per wire were

4.3 Exemplary results

measured and calculated. However, around the dip details of the experimentally determined near-field intensity are not accurately reproduced by the calculated intensity. The differences can be attributed on the one hand to our rough approximation of the true distribution of fields coupling to the guided fibre modes and, on the other hand, to the neglected interactions between probe and sample. These two issues would have to be addressed to further improve our model. A third option is completely independent of the model: the difference between the shapes of the measured and calculated extinction dips suggests some inaccuracy in the first calculation step.

As a last trial for our model, we used the calculated magnetic fields instead of the electric fields to obtain the near-field signal from the single-wire samples. We adjusted the probe model to approximate the magnetic field distributions shown in Fig. 4.1. For this, both H_x and the side-lobes of H_y were neglected and the aperture functions given in Figs. 4.3 (c) and (d) were used, such that

$$H'_x \propto \frac{1}{A}\int_A H_y \, dA + C\Delta_y H_z \quad \text{and} \quad H'_y \propto \frac{1}{A}\int_A H_x \, dA + C\Delta_x H_z. \quad (4.5a)$$

Obviously, this represents a purely phenomenological approach. For a more sophisticated model the magnetic fields and the currents induced in the near-field probe would have to be analysed in greater detail. Nevertheless, as can be seen in the right-hand side column of Fig. 4.5, our coarse approach yields near-field distributions that are extremely similar to the electric-field derived near-field distributions. Note that for this particular structure the wavelength-dependence of the magnetic-field derived signal vanishes. This is discussed in greater detail in chapter 5.2.2.

In conclusion, the calculated images reproduce the general pattern of the measured near-field signal far better than previous attempts made in our group (some of them shown along with the measurements in [46, 78, 90]). For our purposes the approximations for the model seem quite useful. Both here and in the following chapters the computed images agree well with the respective experimental results.

5 Double-wire structure

The first metamaterial design we examined experimentally is the double-wire structure. As detailed in chapter 2.2.2, this structure exhibits a magnetic response at optical frequencies and is an essential component of the more complex fishnet structure (studied in chapter 6). Due to its reduced dimensionality it also allows us to explore the limitations of our simulation model with respect to frequency-dependent calculations.

We begin the chapter with preparatory measurements to relate topographical and optical information (section 5.1). Then we present the main sample and experimental (section 5.2.1), as well as theoretical (section 5.2.2) investigations of its near-field distribution, and discuss the spectral evolution of the near-field signals (section 5.2.3).

5.1 Preparatory examinations

Relating features of the measured near-field images to positions on the sample can present an experimental challenge, because the topographical and optical data obtained during measurements in the constant-distance mode (see chapter 2.4.3) do not necessarily originate from the same position on the sample. As discussed in chapter 4, the optical signal is composed of the near-fields at and around the aperture In contrast, the topographical signal is determined by the lowermost point of the tip. This is usually a small surface asperity on the metal around the aperture, that can be located at a distance of up to 300 nm in any direction from the centre of the aperture. Thus, the positioning uncertainty is 600 nm. Even with an unusually large lattice constant of 1200 nm, the topographical and optical features can not always be matched unambiguously. As the vertical movements of the probe also cause artefacts in the optical signal, we abandoned recording the topography and chose the constant-height mode for our measurements.

To nonetheless obtain positioning information, preparatory measurements at a wavelength of 1310 nm were conducted, beginning with three double-wire samples with the same nominal wire width of 220 nm and lattice constants of 700 nm, 900 nm, and 1200 nm. All double-wire samples were fabricated by Stefan Linden. A sketch of the double-wire structures is shown in Fig. 5.1 (a), along with far-field intensity transmittance spectra of the three samples used for

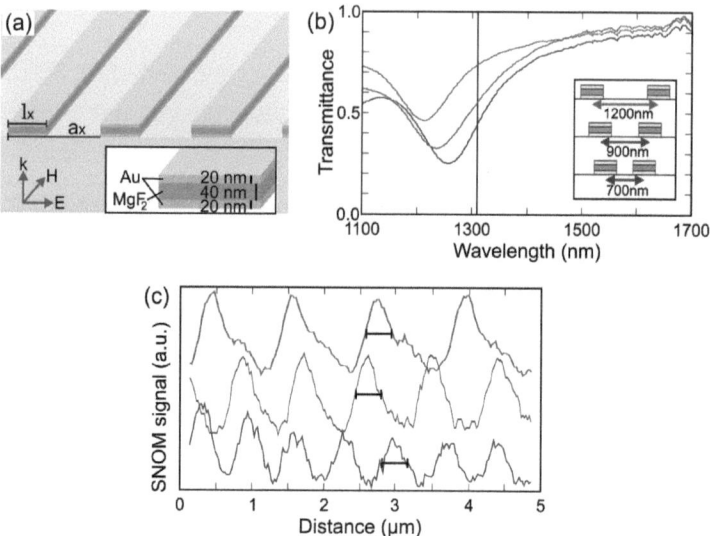

Fig. 5.1: (a) Sketch of the double-wire structures with the incident light wave specified in red. (b) Far-field intensity transmittance spectra. Colour coding as defined in the inset. The nominal width of the wires is 220 nm, the wavelength of the near-field measurements, 1300 nm, is marked by the vertical line. (c) Corresponding near-field intensity, offset for clarity. As guides to the eye, black bars with a width of 350 nm are given at half the maximum of the curves.

the preparatory measurements (Fig. 5.1 (b)). Due to the similar wire widths the resonances of the three samples occur at comparable wavelengths. Thus, the distributions of electromagnetic fields and currents in the wires are expected to be the same, as well. The results of near-field measurements of the three structures are compared in Fig. 5.1 (c) (the black bars serve as guides to the eye). The optical near-field intensity signals from all structures exhibit distinct peaks with a pitch equal to the lattice constant. Due to symmetry the location of these peaks has to be either on top of the wires or between them. As the peak width is clearly roughly the same for all samples, while the peak spacing follows the associated wire spacing, we have to conclude that the peaks occur on top of the wires. Experiments with wire widths between 160 nm and 280 nm, each with the same three lattice constants (not shown), reproduced this finding, as did measurements with specially fabricated protrusion probes with Matteo Burresi (compare with chapter 6.2.1).

Interestingly, this near-field intensity distribution is quite counter-intuitive as the highest signal is found behind the opaque parts of the structure. Here, a naive description by ray optics would predict no signal at all due to shadowing. However, at wavelengths near the magnetic resonance a plasmon is excited on the wires. As discussed in chapter 2.2.2 it generates electric fields at

5.2 Frequency-dependent investigations

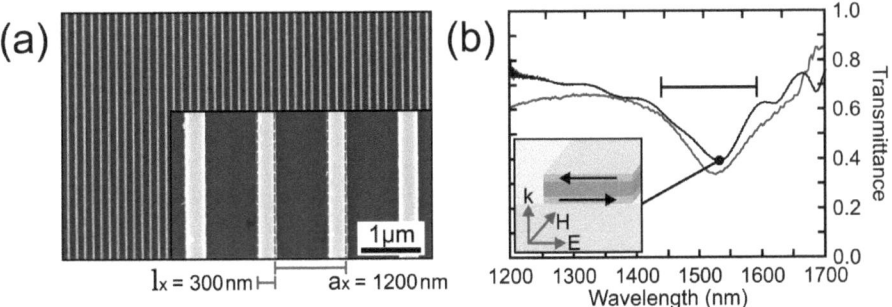

Fig. 5.2: (a) SEM micrographs of a typical area of the double-wire structure with a zoom-in in the inset. The dashed red lines indicate the dimensions of the structure. (b) Measured (blue) and calculated (black) far-field intensity transmittance spectra with the wavelength range of the near-field measurements marked in black. The inset indicates the illuminating wave (red) and the excited mode.

the edges of the wires and a strong magnetic field in the spacer layer between them. The electric fields at the edges have opposite phases and, thus, produce a strong gradient across the aperture ($\Delta_x E_z$, see chapter 4.1), when the probe is above the double-wire. This is the dominating contribution to the measured signal and results in the measured peaks above the wires. To investigate this in greater depth we conducted more detailed experiments on the double-wire structures.

5.2 Frequency-dependent investigations

Spectrally resolved, phase- and polarisation-sensitive near-field measurements of a double-wire structure were taken together with, and by, Matteo Burresi at the group of Prof. Kuipers of the FOM Institute AMOLF (Foundation for Fundamental Research on Matter Institute for Atomic and Molecular Physics) in Amsterdam. SEM images of the sample (with a wire width of 300 nm and a lattice constant of 1200 nm) are shown in Fig. 5.2 (a), corresponding far-field intensity transmittance spectra in Fig. 5.2 (b) (measurement in blue and calculation in black). The resonance is at a wavelength of 1525 nm (1535 nm in the calculation), the inset in (b) illustrates the corresponding excited mode and the polarisation direction, with the electrical field perpendicular to the wires. For the calculation the parameters shown in (a) were used, along with a height of the gold and MgF_2 of 20 nm and 40 nm, respectively. Measurement and calculation agree well within the spectral region of the resonance, where the main near-field measurements were taken (wavelength range indicated in black).

5 Double-wire structure

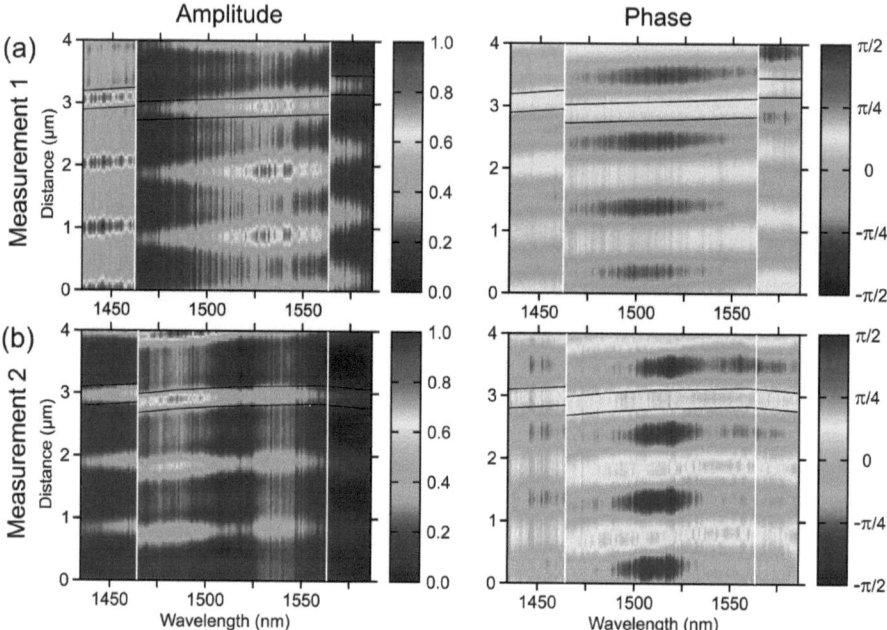

Fig. 5.3: Normalised near-field amplitude and phase distribution of two measurements—(a) and (b), respectively—versus wavelength. The position of one double-wire is marked in black. Discontinuities at the white lines are due to realignment. The associated sample is pictured in Fig. 5.2.

5.2.1 Near-field measurements

The probes used for the SNOM measurements were coated with 150–200 nm aluminium and had nominal aperture diameters of 200 nm. Due to the symmetry of the samples, no signal (except from imperfections in probe and sample) is expected in the secondary polarisation channel, that is, parallel to the wires. Accordingly, the near-field signal in this channel was much lower than that in the main polarisation channel (perpendicular to the wires), namely by a factor of 8.2 in the first measurement and 5.3 in the second measurement. Near-field amplitude and phase images from the main channel are shown in Figs. 5.3 (a) and (b) (first and second measurement, respectively). Due to lack of an optical reference signal, the amplitude of each image is normalised to its global maximum and the phase for each wavelength is given as the deviation from the phase of the complex average of the near-field signal. On account of realignments that become necessary if large wavelength ranges are covered, the data were collected in several instalments, causing some discontinuities (indicated in white). The location of one double-wire, as inferred from the preparatory measurements, is sketched in black. While for the single-wire structures in chapter 4.3 a contrast reversal at shorter wavelengths was observed, the double-wires show no such inversion. Instead, in the entire wavelength range examined, the amplitude of the signal is higher above the wires than between them. Similarly, the phase above the wires shows a uniform phase delay (*i.e.*, a higher phase) with respect to that between the wires.

5.2.2 Calculations

We now compare the measured near-field distributions to calculated images. The aperture was modelled at a height of 10 nm above the sample with a complex factor of $C = 2.7 \cdot e^{-i \cdot 0.12\pi}$ and a diameter of 280 nm. This aperture diameter is rather larger than that found by SEM for the probe used in the measurements. However, due to the finite penetration depth of light in metal, optical apertures of near-field probes are always somewhat wider than seen in the SEM (see chapter 3.1.2). Furthermore, as discussed in chapter 4.2.2, larger aperture diameters are expected for our model.

The resulting simulated near-field amplitude and phase distributions for the double-wire structure are displayed in Fig. 5.4 (a) in the same manner as the measured distributions in Fig. 5.3. Again, the amplitude signal is highest above the wires and lowest between them. As above, it is dominated by the gradient contribution ($\Delta_x E_z$, see chapter 4.1). We attribute the narrow amplitude dip at the centre of the wire to the simplified aperture functions of the probe model. The phase, too is highest above the wires. However, it exhibits a reversal at longer wavelengths, which is not seen in the experiment. This is a result of an issue in our model which we have already remarked upon (chapter 4.2.2), namely the frequency dependence of C. Here, C is phenomenologically chosen to produce good results at a wavelength of 1500 nm. If C changes significantly with wavelength, the results of the calculation at other wavelengths than 1500 nm will deviate from the experimental data. Fortunately, instead of having to adjust C phenomenologically for every wavelength used in the calculations, we can take advantage of the reduced dimensionality of the double-wire structure.

Magnetic field calculation

Recall, that while the wires on the sample are 300 nm wide in the x-direction, they are 100 µm long in the y-direction. For all practical purposes they can, therefore, be considered as infinitely long. Due to symmetry, E_y and all partial derivatives in y—that is, $\partial_y E_x$, $\partial_y E_y$ and $\partial_y E_z$—then vanish. Hence, $H_x = \partial_y E_z - \partial_z E_y = 0$ and $H_z = \partial_x E_y - \partial_y E_x = 0$. Following the reasoning discussed in chapter 4, the simulated SNOM signal derived from the magnetic fields then becomes proportional to H_y in the aperture and, therefore, independent of C (compare with Eq. 4.5a). The reduced dimensionality of the double-wire structure thus allows us to

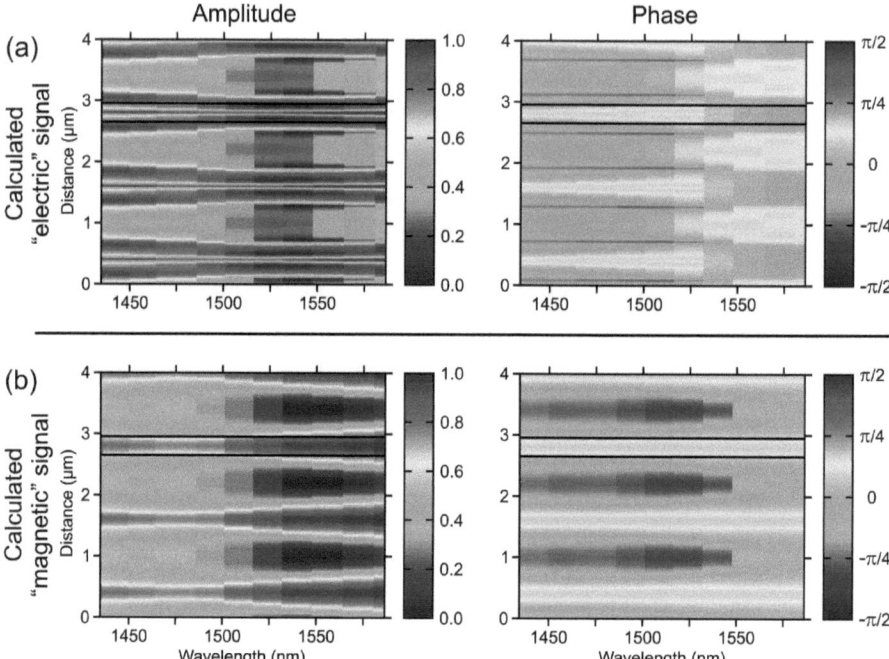

Fig. 5.4: Calculated near-field amplitude and phase distributions derived from (a) the electric fields and (b) the magnetic fields. The position of a double-wire is marked in black. Details of the calculations are given in the main text.

calculate the near-field signal for different frequencies without introducing additional simulation parameters.

The magnetic-field derived near-field signal (amplitude and phase) is depicted in Fig. 5.4 (b). Obviously, the agreement with the measurements in Fig. 5.3 is far better than that of the electric-field derived signal. The amplitude distribution is reproduced without the inaccurate dip above the centre of the wire, and the phase image is even quantitatively an extremely good replica of the measured phase distribution, exhibiting nearly the same peak width above the wires and a very similar dip in the frequency evolution between the wires.

5.2.3 Spectral behaviour

The spectral evolution of the amplitude and phase signal can be seen more clearly in Fig. 5.5, where the difference between the maximum and minimum value of both amplitude and phase

5.2 Frequency-dependent investigations

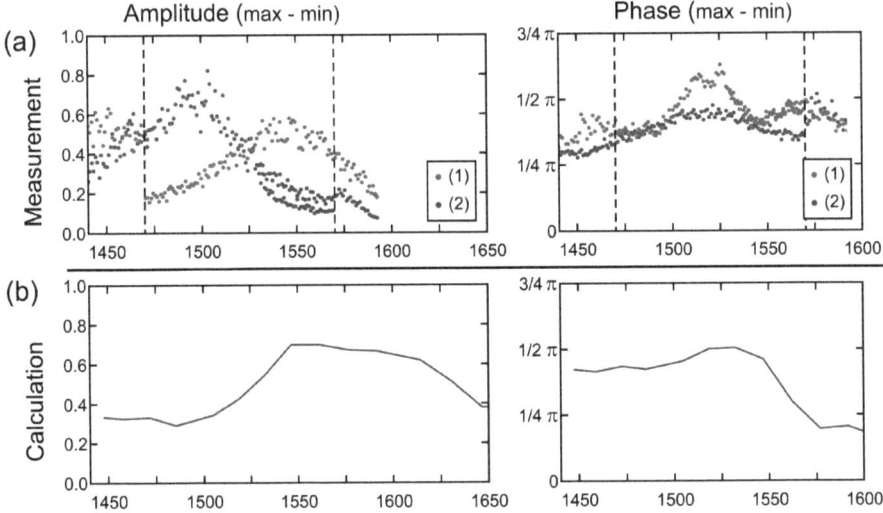

Fig. 5.5: Difference between maximum and minimum amplitude and phase versus wavelength, (a) measurements and (b) calculation. Discontinuities at the dashed lines are due to realignment.

is plotted versus wavelength. Here, again, the overall agreement between experiment and calculation is surprisingly good.

In the amplitude the exact spectral position of the peak apparently depends on the probe used. This could be an effect of slightly different probe geometries, which might cause some wavelengths to scatter more efficiently into the aperture. Alternatively, probe-sample interactions might induce an energy shift in the double-wires' resonance that depends on the scanning height. Further investigations within a greater range of wavelengths, with a larger variety of near-field probes, and especially at different scanning heights to influence the interaction strength might help to elucidate this issue.

The relative phase delay in Fig. 5.5, both in the measurements and the calculation, has its peak approximately at the resonance wavelength of the double-wires (1525 nm in the measurements, 1535 nm in the calculation). The maximum relative delay is roughly 0.5π. Here, again, variations between the two experiments and between experiments and theory have to be attributed to differences between the near-field probes and scanning heights.

6 Fishnet structure

After the investigation of double-wire structures in the previous chapter we now turn to the composite structure made up of double-wires and a diluted metal, the double-layer fishnet. The far-field properties of this fishnet design are well studied [49, 50, 91, 92]. Among them, as discussed in section 2.2.3, is the combination of negative $\varepsilon_r^{\text{eff}}$ and μ_r^{eff} that results in the negative refractive index of the fishnet. This negative index is central to the many effects mentioned at the beginning of chapter 2.2. As most of them operate at the nanoscale, not only the effective material parameters of the negative-index structure have to be taken into account for their implementation, but also the structure's near-field behaviour.

In order to further our understanding of this nanoscale behaviour, phase- and polarisation-sensitive near-field measurements of a fishnet structure were taken by Matteo Burresi at the group of Prof. Kuipers [79]. We begin the chapter with the far-field properties of the structure (section 6.1). Then, we discuss the near-field's amplitude and phase distribution in the immediate vicinity of the fishnet (section 6.2.1) and show its spectral evolution (section 6.2.2), as well as the transition from near-field to far-field (section 6.2.3).

6.1 Far-field properties

For the measurements a fishnet sample fabricated by Gunnar Dolling was used. A sketch of it, as well as an SEM image, is shown in Fig. 6.1. Far-field intensity transmittance and reflectance

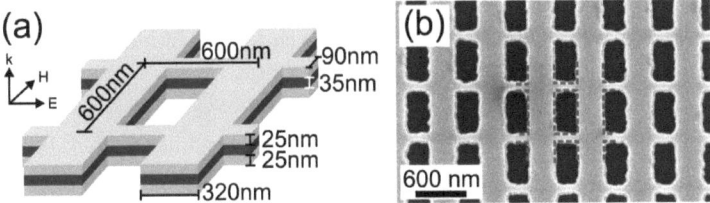

Fig. 6.1: (a) Sketch of the fishnet structure with dimensions as used in the calculations, except for the film thicknesses. Adapted from [92]. (b) SEM micrographs of a typical area of the fishnet sample [79]. The dashed red lines illustrate the dimensions given in (a).

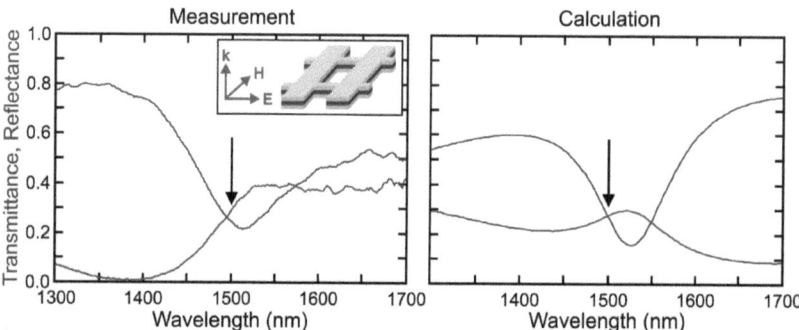

Fig. 6.2: Measured (a) and calculated (b) far-field intensity transmittance and reflectance spectra of the fishnet structure shown in Fig. 6.1. Arrows indicate the wavelength of the main near-field experiments.

spectra as measured by Gunnar Dolling are given in Fig. 6.2, along with calculated spectra. The dimensions shown in Fig. 6.1 were used in the calculations, except for the thicknesses of the gold and MgF$_2$ films, for which we used 30 nm and 40 nm, respectively, in order to reproduce the far-field properties of the sample more closely. Measurement and calculation agree fairly well within the spectral region of the resonance, where most of the near-field measurements were taken (1470 nm to 1600 nm). The wavelength of the main near-field experiment, 1500 nm, is indicated by arrows. It lies both within the resonance dip and inside the region of the known negative refractive index of identical fishnet structures.

6.2 Near-field investigations

6.2.1 Near-field distribution

In order to be able to relate features of the measured near-field images to positions on the fishnet sample, preparatory measurements were conducted, for which near-field probes with a dielectric protrusion (typically 100–200 nm) were fabricated. As the apex of this protrusion is located approximately below the centre of the aperture, the topographical details detected with it originate from the same locations on the sample as the optical signal picked up through the aperture. Thus, topographical and optical signals coincide spatially. Results of the preliminary measurements at a wavelength of 1500 nm are displayed in Fig. 6.3, namely the topography (a), in arbitrary units, as well as amplitude (b) and phase (c) of the near-field signal. As in chapter 5 the amplitude is normalised to its maximum and the phase is given as the deviation from its average. The positions of the holes in the fishnet (marked in black) can easily be inferred from

6.2 Near-field investigations

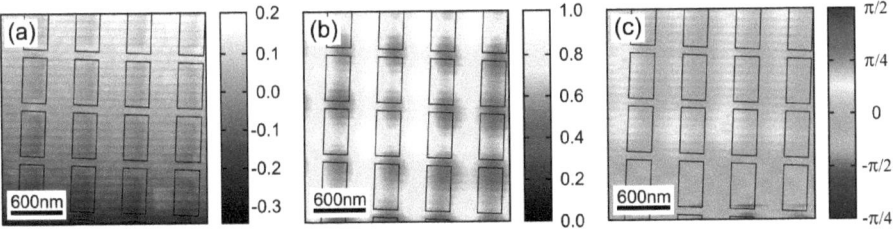

Fig. 6.3: Topography (a) of the fishnet sample, as well as the amplitude (b) and phase (c) of its near-field distribution at a wavelength of 1500 nm. Measured with a near-field probe with a dielectric protrusion. [79]

the topographical image. Apart from a slight lateral offset (approximately 6% of the length of the unit cell) they fit very well into the amplitude and phase images. We can transfer the information gained from this measurement directly to the near-field images measured with a conventional probe, provided that the disturbance of the optical signal caused by the glass protrusion is small.

This is indeed the case, as can be seen from the near-field optical images of the fishnet structure, presented in the top row of Fig. 6.4, which were taken with a conventional probe at the same wavelength in the constant-height mode. Again, the near-field signal is normalised to the maximum of the amplitude and the mean value of the phase. The agreement between the near-field distributions measured by the conventional probe and by the protrusion probe is quite obvious: in both measurements the amplitude as well as the phase distributions are extremely similar. Note, that the variation of the phase picked up by the protrusion probe is smaller than that measured by the conventional probe. As the protrusion keeps the probe at a greater distance from the sample, this is to be expected (discussed below in greater detail, see section 6.2.3). Similarly, the amplitude variation is less extensive than that in the conventional images. Due to the otherwise strong agreement we can be confident in assigning the positions of the holes to the conventionally taken measurements (again indicated in black in Fig. 6.4). As observed with the double-wire structure, the amplitude and phase signals are both highest above the thick wires (the "double-wires" of the fishnet), while they are much lower between them, i.e., above the holes and the thin wires (the "diluted metal").

The calculated near-field distribution 10 nm above the fishnet structure, also at a wavelength of 1500 nm, is depicted in the bottom row of Fig. 6.4, on the same scales and with the same normalisation as the measured field distribution in the top row. As for the probe used on the double-wire structure, we modelled the aperture with a diameter of 280 nm and a complex factor of $C = 2.7 \cdot e^{-i \cdot 0.12\pi}$. Obviously, the agreement with the measured images is very good. The location of the amplitude peaks is reproduced perfectly and the signal exhibits a phase lag

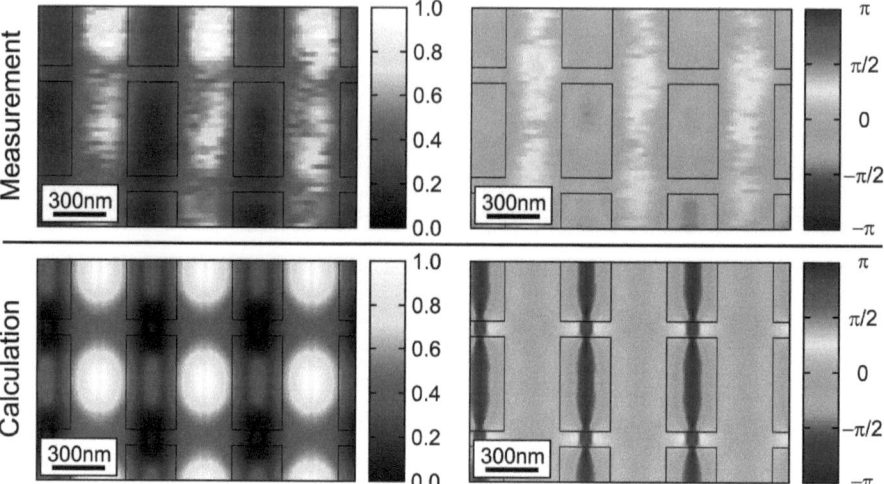

Fig. 6.4: Normalised measured and calculated near-field amplitude and phase distribution for a wavelength of 1500 nm at a height of 10 nm above the fishnet shown in Fig. 6.1. [79] The locations of the fishnet holes are indicated in black.

above the thick wires. Again, small differences can be attributed to the approximations in our model.

6.2.2 Spectral behaviour

Similar to our observations on the double-wire structures in chapter 5, the amplitude and phase distributions of the measured near-field signal above the fishnet structure change only slightly in the wavelength range covered by the experiments. This is shown exemplarily for the phase in Figs. 6.5 (a) and (b), which summarise the measured and calculated spectral evolution of the phase at the four high-symmetry points of the structure: the centre of the holes, the centres of both the thick and the thin wires, and the position where the wires cross (see inset). In the measured data the general appearance of the distribution remains unchanged. Only the difference between the phase above and between the thick wires decreases from approximately $\pi/2$ at a wavelength of 1470 nm to roughly $\pi/4$ at 1590 nm. The calculated signal reproduces this phase only for wavelengths around 1500 nm, for which the complex factor C was specified. Particularly at longer wavelengths discrepancies evolve. Unfortunately, unlike in chapter 5.2.2 we cannot disregard C for the magnetic-field derived near-field distributions, because the fishnet structure does not possess the double-wires' translational symmetry.

6.2 Near-field investigations

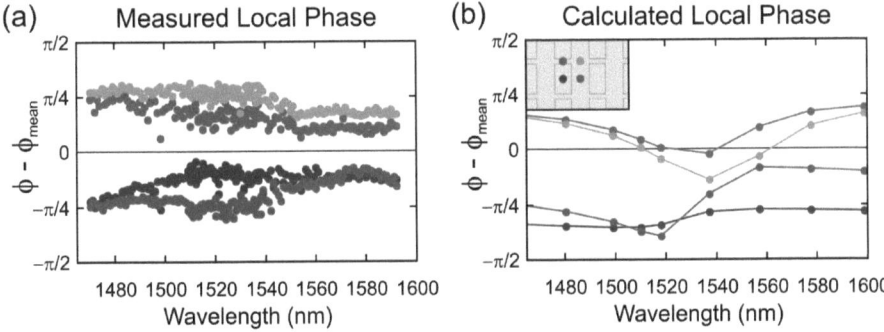

Fig. 6.5: (a) Measured and (b) calculated spectral evolution of the phase above the four high-symmetry points of the fishnet structure (see inset). For each wavelength the phase is referenced to the mean phase. [79]

Previous measurements by Gunnar Dolling on a sample identical in construction [91] showed that the far-field phase delay caused by the fishnet sample takes on values from $\approx -0.12\pi$ to $\approx +0.12\pi$ in this range of wavelengths (corresponding to a refractive index change from -1 to 1). The variation observed in the near-field strongly exceeds this far-field value. It is interesting to compare the two in greater detail. Fig. 6.6 shows a histogram of the phase distribution versus wavelength. To facilitate the comparison, the near-field phase is referenced to the known far-field delay (shown in red). Each slot shows the unit-cell-integrated amplitude of data points with the corresponding phase, *i.e.*, the "amount" of light with that phase. The colour scale runs from white (no signal) to black (maximum of the histogram). Again we see that the near-field's phase variation within a unit cell far exceeds the variation in the far-field delay. Thus, at this distance the fishnet cannot be described as a homogeneous medium. This has to be taken into account in the design of potential nanoscale applications of the negative-index fishnet structure—such as the perfect lens—which operate within the extension of the near-field.

6.2.3 Transition to the far-field

In order to determine to what distances the phase variations extend, the SNOM signal at a wavelength of 1500 nm was determined experimentally for different heights above the sample. As the wavelength was fixed in these measurements, corresponding calculations could be executed. Histograms of the phase variation within the unit cell versus height of the near-field probe are given in Fig. 6.7, derived from (a) experimental and (b) theoretical data. Both measurement and calculation show a pronounced decrease in the phase variations within a distance of approximately 250 nm (in the calculation the variations decrease marginally more slowly than in the experiment, which we attribute to the larger aperture). Only at distances larger

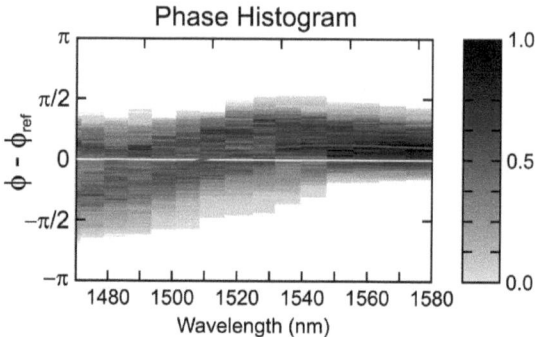

Fig. 6.6: Histogram of the measured near-field phase above the fishnet sample, referenced to the known far-field phase delay (shown in red).

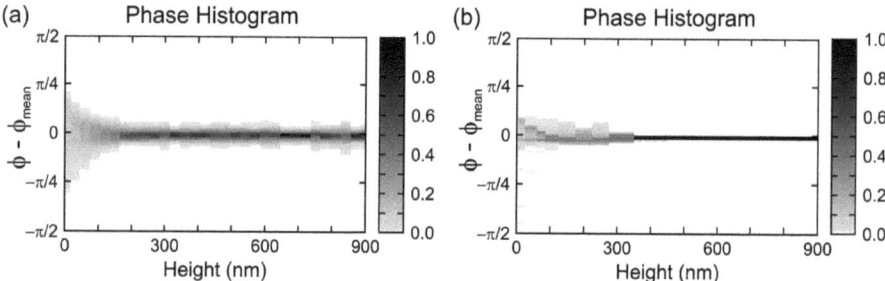

Fig. 6.7: Histograms of the (a) measured and (b) calculated near-field phase at different heights above the fishnet sample. [79]

than this the phase variations are sufficiently small to approximate the transmitted light by a plane wave. Then, we can consider the fishnet structure as a homogeneous material and assign effective parameters. However, for distances below it the near-field effects cannot be neglected and a description of the metamaterial by effective material parameters, $\varepsilon_r^{\text{eff}}$ and μ_r^{eff}, fails. Thus, we find the transition between the near-field domain and the far-field domain at approximately 250 nm distance from the sample.

7 Low-symmetry split-ring-resonator arrays

In this chapter, we investigate the near-field properties of low-symmetry split-ring-resonator arrays [80]. As explained in chapter 2.2.4, these structures consist of pairwise rotated split-ring resonators whose magnetic moments interact, causing a splitting of the fundamental SRR resonance into two eigenmodes: one with parallel magnetic moments (high-energy eigenmode) and one with antiparallel magnetic moments (low-energy eigenmode) [3]. As the electric coupling between neighbouring SRRs is expected to be negligible (because neighbouring electric moments are perpendicular), any effects of interactions in the low-symmetry arrays can be attributed to the magnetic coupling only. Thus, they are ideally suited to investigate and visualise the effects of magnetic interactions between the individual split-rings.

We begin our study by examining the requirements imposed on the structures by the near-field experiments, and show SEM data and optical far-field characterisations (chapter 7.1). Then, measured near-field images for the eigenpolarisations (at $\pm 45°$) are presented and discussed, as well as the images for horizontal and vertical polarisation (chapter 7.2). As before, we compare all experimental results to calculated near-field distributions that reproduce the measurements and support our conclusions.

7.1 Far-field properties

The planar, low-symmetry split-ring-resonator arrays ("l-SRR" arrays) used in our measurements were manufactured by Manuel Decker. For the purpose of control experiments a "complementary" structure, that is, an identical l-SRR array rotated by 90 degrees, was produced on the same sample as the original structure. Additionally, an array with the usual, uniform orientation of the SRRs and its rotated counterpart were generated.

In order to ensure reproducibility of the obtained near-field images, special attention was given in the fabrication to making the arrays as reliably periodic and free of irregularities as possible. In particular, nanoscale impurities on the samples, which might be picked up by the probe and alter its transmission properties, were avoided. Prior to the experiments we inspected

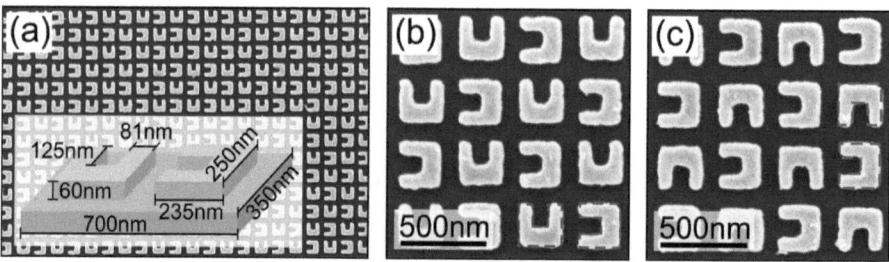

Fig. 7.1: SEM micrographs of (a) a typical area of the low-symmetry SRR array (inset is a sketch of the array's unit cell), (b) a zoom-in of (a) with a single unit cell marked in red according to the dimensions given in (a), and (c) a zoom-in of a 90° rotated array on the same substrate with the unit cell marked in green. [80]

the samples with an SEM to verified their quality and obtain structure parameters. Resulting micrographs of the l-SRR structures and the parameters are shown in Fig. 7.1. The inset indicates the dimensions of the individual resonators and the unit cell. They were chosen to satisfy several conditions:

Most importantly, the SRRs have to be considerably larger than the minimum resolution of our near-field microscope. Enlarging the resonators, however, reduces the resonance frequencies of the SRRs for the two eigenpolarisations, which have to stay within the range of wavelengths available to the SNOM. Fortunately, increasing the width of the split-ring's arms raises the resonance frequency and thus to a certain extent counterbalances the redshift due to the enlargement of the rings. Ultimately, the SRRs were sized to resonate at wavelengths of 1465 nm and 1550 nm (determined with far-field transmittance measurements), such that at 1510 nm the transmittance for the two eigenpolarisations is approximately equal. This means that at 1510 nm the "ferromagnetic" and the "antiferromagnetic" mode are both excited with similar strength. Yet another matter to take into account is that the magnetic coupling and the effects resulting from it are expected to be larger when the split-rings lie close to each other. Therefore, the spacing between the individual resonators should be small compared with their size. This has an additional benefit for scanning in the constant distance mode: the SRRs are placed too closely for the near-field probe to dip down between them. Thus, artefacts related to changes in probe height are avoided. On the other hand, with smaller resonator distances the overall transmittance of the array is reduced, along with the expected near-field intensity. Using a thinner gold film partly compensates for this, but also red-shifts the resonance. Consequently, the signal intensity has to be enhanced by employing larger apertures with higher transmittance but lower resolution. This leads to a trade-off between strong coupling and high resolution.

As a result of these considerations we specified the dimensions of the split-ring arrays in Fig. 7.1 as the optimum for our near-field investigations. Measured far-field transmittance spectra of

7.1 Far-field properties

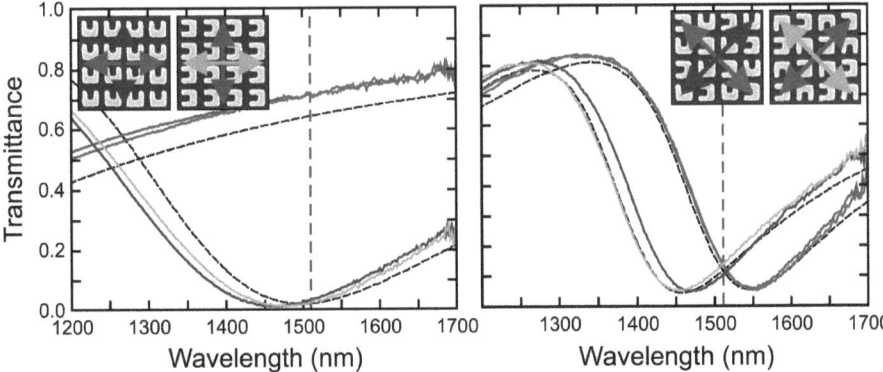

Fig. 7.2: Measured far-field transmittance spectra of the low-symmetry SRR arrays (left-hand side) and of usual SRR arrays (right-hand side). Polarisation directions as defined in the insets. Dashed black curves are calculated spectra. The vertical grey lines mark the wavelength of the near-field measurements. [80]

the pictured samples are shown on the right-hand side of Fig. 7.2. For comparison, we display spectra of the usual (uniformly oriented) SRR arrays on the left-hand side. The dashed black curves are calculated data (Microwave Studio, dimensions as given in Fig. 7.1). As described in chapter 2.2.1, the fundamental mode of the uniformly oriented split-rings is only excited when the incident polarisation is parallel to their electric moment, that is, horizontal for the originally oriented SRRs (gap pointing up) and vertical for the 90° rotated ones (gap pointing sideways). No resonance appears for the orthogonal polarisations in this range of wavelengths. As expected, the response of the structure rotated by 90 degrees is the same as that of the original one if the illuminating polarisation is rotated by the same angle. Obviously, the agreement between the spectra of the complementary structures is nearly perfect, indicating extremely low fabricational asymmetries. Furthermore, the calculated data also fit the measured curves well. The same holds true for the l-SRR arrays (right-hand side of Fig. 7.2): here, too, the agreement between the complementary structures is surprisingly accurate and the calculated data fit perfectly. Note, that the magnetic resonance splits into the higher-energy "ferromagnetic" (blue and light blue) and the lower-energy "antiferromagnetic" mode (red and green), with a large separation of around 90 nm (12 THz). This corresponds to a splitting of 6% of the centre frequency, indicating rather strong coupling for this type of structure [3]. Note also, that at the wavelength used for the near-field examinations (vertical grey line) the transmittances of the two modes match.

7.2 Near-field investigations

7.2.1 Diagonal polarisation

For the measurements we chose the constant-distance feedback-control mode, as it compensates for drifts between sample and probe. Importantly, as mentioned before, variations in the height profile of our l-SRR arrays occur on length scales far smaller than the probe tip. Thus, the probe is in fact scanned at a constant height and no artefacts related to changes in probe height have to be expected. This is confirmed by the topographical images taken simultaneously to the near-field images (not shown).

Selected results of the near-field measurements are depicted in Figs. 7.3 (a) to (d), and compared to calculated images (e). The incident light has a wavelength of 1510 nm and diagonal linear polarisations of $-45°$ (left-hand side column) and $+45°$ (right-hand side column), as indicated at the top. These are the two eigenpolarisations of the l-SRR arrays. All ten images in Fig. 7.3 are represented on the same spatial scale, which is given at the top of the figure. Similarly, the colour scale is identical for all images, measured as well as calculated. The near-field intensity signal is normalised to the near-field intensity above the bare glass substrate (reference measurement and reference calculation, respectively). This allows a direct and, most notably, quantitative comparison between experimental and theoretical data.

Let us begin by discussing the first row of Fig. 7.3 (a). Shown here are measured near-field distributions of the original, non-rotated low-symmetry SRR array. For the eigenpolarisation associated with the "antiferromagnetic" mode (left-hand side) we observed bright diagonal bands, whereas separated and much dimmer spots were obtained for the "ferromagnetic" mode (right-hand side). Recall, that without coupling between the SRRs, the two images should look alike. The pronounced difference between them is, therefore, a direct consequence of the in-plane coupling between the split-ring resonators in the array.

Of course, in principle, differences between the images could also originate from asymmetries of the l-SRR array. In order to eliminate this possibility, control experiments were carried out. In Fig. 7.3 (b) equivalent measurements were conducted on the complementary array (an identical array rotated by 90 degrees). As expected, the measured distributions are rotated with respect to the earlier measurements and the "antiferromagnetic" mode is now excited by the $+45°$ polarisation and the "ferromagnetic" mode by $-45°$. Thus, the measurement results agree extremely well. To further rule out asymmetries in the near-field probe used for the measurements, the entire substrate was then rotated by 90 degrees and the measurements were repeated (shown in Figs. 7.3 (c) and (d)). Again, excellent agreement was found.

In the last row of Fig. 7.3 we compare the experimental results to calculated images at a wavelength of 1510 nm and a height of 20 nm above the sample. We implemented the probe model

7.2 Near-field investigations

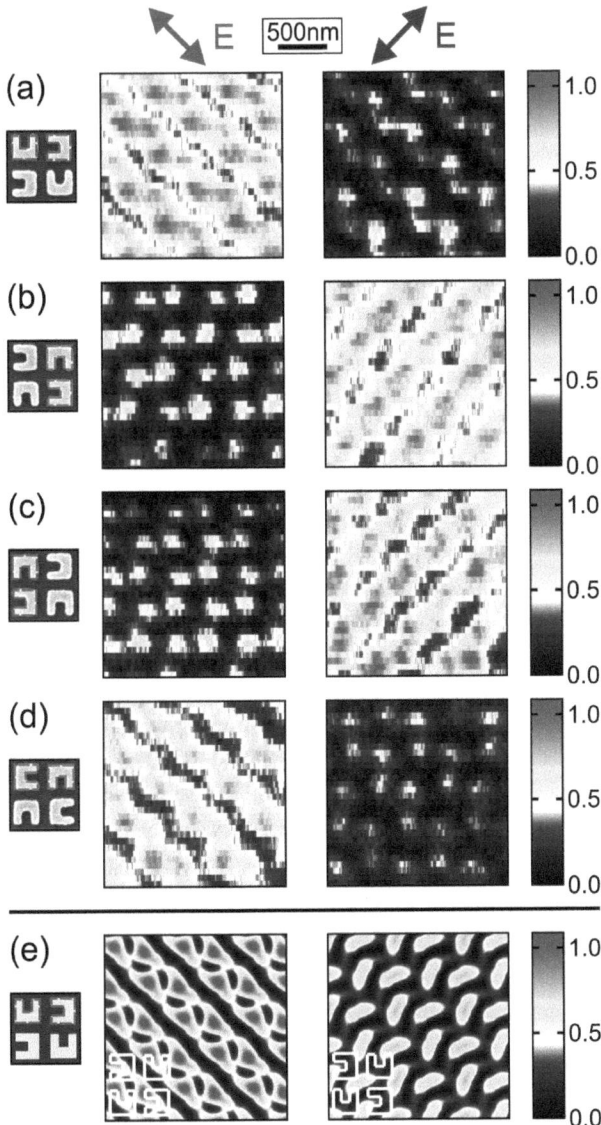

Fig. 7.3: Measured near-field images, (a) to (d), and calculated ones (e) of the structures indicated on the left. For (c) and (d) the samples of (a) and (b) were physically rotated by 90°. Polarisation directions are shown at the top. All images, including the SEM micrographs, are depicted on the same scale (given at the top). The colour scale is normalised to the near-field intensity over a glass substrate. [80]

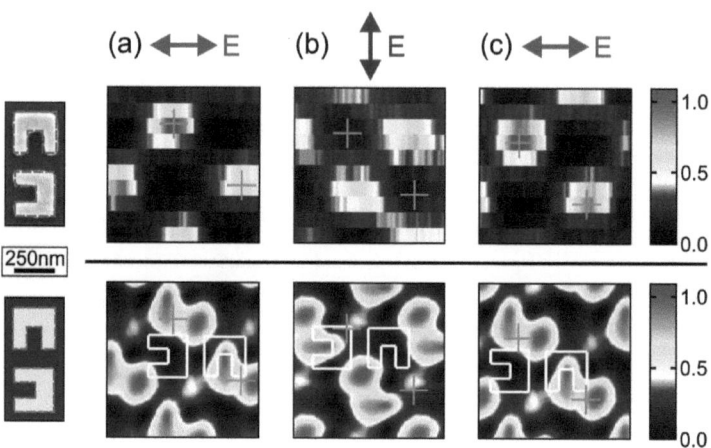

Fig. 7.4: Near-field images (top row) and calculated images (bottom row) of an l-SRR structure under illumination with horizontal (a) and (c) and vertical (b) polarisation. All images are normalised to the respective reference. [80]

with an aperture diameter of 280 nm and a complex factor of $C = 1.6 \cdot e^{i \cdot 0.35\pi}$, and referenced the results to the simulated signal over a bare substrate. The position of the resonators in the calculated distribution is indicated in white. Consistent with the measurements we once again obtained bright, connected bands for the "antiferromagnetic" mode, whereas the calculations yielded separated and dimmer spots for the "ferromagnetic" mode. Note, that the results are even quantitatively comparable to the measurements. Quite obviously, experiment and theory fit extremely well. Thus, the influence of probe–sample interactions on the near-field images is far weaker than that of the magnetic interactions between the individual resonators. Again, this reassures us that the probe's influence can indeed be neglected in our model.

7.2.2 Horizontal and vertical polarisation

Whereas we found a pronounced difference between the near-field images of the two eigenmodes, the distributions for the horizontal and vertical polarisation directions have a very similar structure. In the top row of Fig. 7.4 we display images taken with horizontal and vertical polarisation. The small drop in the amplitude of (b) can be attributed to the alignment of the laser spot onto the probe. Otherwise the images show a good overall resemblance, in that they feature well separated, bright spots at regular intervals. In the bottom row calculated images are depicted for comparison. Again all images are on the same spatial scale (given on the left) and colour scale (on the right). Most strikingly, the images exhibit only one bright spot for every unit cell of the l-SRR structure, whereas two were observed for the diagonal polarisations

7.2 Near-field investigations

(Fig. 7.3). Recall, that every unit cell is made up of two split-ring resonators, only one of which can be excited directly with horizontal polarisation, while the other can only be excited with vertical polarisation at 1510 nm. In order to determine whether the spots can be associated with the respective excited resonators, we compare the three images in the top row of Fig. 7.4. They were taken sequentially on the same sample area under illumination with (a) horizontal, (b) vertical, and (c) again horizontal polarisation. The two central spots in (a) and (c) are marked with green crosses. Due to a slight lateral drift those in (c) are offset with respect to those in (a) by approximately 190 nm. The crosses in (b) are placed midway between the positions of those in (a) and (c). Clearly, they mark the darkest areas in (b). Thus, when the polarisation is rotated from horizontal to vertical, the brightest and darkest spots switch places, indicating the aforementioned association that different SRRs are excited for each polarisation.

Yet again, the calculated images (bottom row of Fig. 7.4) reproduce the measured near-field distributions quite well. In full accordance with the measurements we observe well separated spots and the same switch between the positions of the brightest and darkest areas for the polarisation rotation. In contrast to the measurements, the positions of the individual split-ring resonators are known in the calculations. They are indicated in white in Fig. 7.4. Obviously, a bright spot is obtained around the gap of each excited resonator, along with a strong extension pointing towards the neighbouring resonator. Instead of this twin peak, the measurements show single, smaller peaks. Again, we attribute the differences between measurement and calculation to the approximations in our model and to interactions between probe and sample.

8 Conclusion

The aim of this work was to study the complex interplay of electromagnetic fields and currents that generates the unusual properties of metamaterials. For this purpose the excited modes in these metallic nanostructures were investigated with a near-field optical microscope (SNOM). We measured the near-field distributions above several metamaterial designs (the double-wire structure, the fishnet structure, and low-symmetry split-ring-resonator arrays) with an aperture SNOM and compared them to data calculated with a specifically constructed model of the SNOM's imaging process.

The development of a numerical model for the near-field probe was an important aspect of this thesis, as complete, three-dimensional field calculations are far too time-consuming to be practical. We described the imaging process theoretically, modelling the probe by a transfer function. For this, interactions between probe and sample were ignored. Instead, the near-fields above the metamaterial structures were regarded as an uninfluenced source, coupled by the near-field probe's aperture into the waveguide modes of the probe's optical fibre. Thus, the near-fields could be computed separately. We employed a commercially available calculation software, "CST MICROWAVE STUDIO" (by CST AG), for the double-wire, fishnet and split-ring-resonator samples, and an existing code, home-made with Matlab (by The MathWorks, Inc.), for the single-wire structures. Utilising the calculated amplitude and phase of the near-fields above the sample, the SMOM images could be obtained by numerically mimicking the probe's transmission characteristics with a complex, vectorial transfer function which was motivated both with physical principles and through numerical calculations. For each polarisation direction it consists of two electric contributions and a complex coefficient C which accounts for the phase and amplitude difference between them. C depends on the geometry of the near-field probe and the wavelength used. It was determined phenomenologically, that is, by optimisation of the agreement between experiment and theory for a fixed wavelength. As an alternative to the electric field components, the magnetic field components can be used for the derivation of the near-field signal. Numerical studies confirming all of these considerations were shown and were used to discuss limits of the model that are due to simplifications. Also, exemplarily, results obtained with our model were compared to near-field measurements on single-wire structures conducted earlier in our group.

After this, we experimentally examined double-wire structures as a test-bed for more complex metamaterials. They were also taken as a basis to further explore the limitations and the advantages of our model. Counter-intuitively, the signal intensity was found to be highest above the opaque parts of the structure, *i.e.*, above the wires. The location of the optical peaks with respect to the wires was established by comparing experimental data from several structures possessing fixed wire widths but different lattice constants. Spectrally resolved near-field measurements of the amplitude and phase distributions revealed a phase delay behind the wires and only small changes with wavelength. The experimental data could be reproduced in spectrally resolved calculations of the magnetic-field derived near-field signal, for which we utilised the reduced dimensionality of the double-wire structure in order to eliminate the wavelength-dependence of the phenomenological parameter C.

Experiments on a fishnet structure were discussed next. Here, we investigated the transition from the strong phase variations in the near-fields at the sample surface to the plane, transmitted waves in the far-field. In order to relate features of the measured near-field images to positions on the sample, preparatory measurements were conducted with a specifically fabricated near-field probe exhibiting a dielectric protrusion below the aperture. Due to the protrusion, the topography of the sample could be detected along with the optical signal. A comparison with near-field images collected at the same wavelength with a conventional probe showed, that the signal disturbance caused by the protrusion was small enough to allow for a transfer of the topographical information to the conventional images. Thereby we determined that the highest amplitude and phase occurs above the thick wires, namely the double-wire constituent of the fishnet. Calculated data reproduced these findings. In spectrally resolved experiments we found strong phase variations within the unit cell, which far exceeded the modulus of the known, negative far-field phase delay caused by the negative refractive index of fishnet structures. Therefore, in the near-field, the fishnet cannot be described by effective material parameters. In the far-field, however, this is routinely done. To determine the distance of the transition between the near-field and the far-field domain, the development of the phase variations with distance was determined experimentally at a wavelength of 1500 nm and was compared to corresponding calculations. Both exhibited a decay of the variations within a distance of roughly 250 nm. At distances larger than this the phase variations were sufficiently small to approximate the transmitted light by a plane wave and consider the fishnet structure as a homogeneous material with effective material parameters.

Finally, we visualised and analysed the effects of magnetic interactions between individual resonators of low-symmetry split-ring-resonator (l-SRR) arrays. We discussed the requirements for the l-SRR samples and presented far-field intensity transmittance spectra. For horizontal and vertical polarisation these showed the same resonance, while for the diagonal polarisation directions they featured a frequency splitting between the (higher-energy) symmetric and the (lower-energy) antisymmetric eigenmode. This splitting amounted to 6% of the centre frequency of the uniformly oriented split-rings, demonstrating strong coupling within the arrays. The coupling effects were then directly visualised by measuring the distinctly different near-field intensity distributions of the two eigenmodes at a wavelength of 1510 nm, directly between the symmetric and antisymmetric resonances. We observed bright, connected bands for the antisymmetric mode and separated, dimmer spots for the symmetric one. With control experiments we eliminated the possibility that these effects originated from either fabricational asymmetries of the samples or imperfections in the near-field probe. Furthermore, a strong, qualitative and quantitative agreement between experimental data and calculated near-field images was found. From this we deduced that any interactions between near-field probe and l-SRR sample, neglected in the theoretical model, have a far weaker effect on the signal distribution than the magnetic interactions between the individual split-ring resonators. Additional measurements under illumination with horizontal and vertical polarisation were compared to corresponding calculations. By these polarisations only every second resonator is excited directly. In both experiment and theory we obtained bright spots, distinctly separated by darker areas. In contrast to the symmetric eigenmode, where the near-field signal exhibited two spots for every unit cell (and, thus, the same number of spots as resonators), here only one spot was observed. Images measured in quick succession with orthogonal polarisation directions revealed a switch in the positions of the dark areas and the bright spots, indicating an association between the spots and the respective excited resonators. Yet again, the calculated intensity signal supported our conclusions.

We introduced our model as a useful tool to reproduce measurement results of near-field optical experiments with aperture probes on metamaterials. In future works its precision might be enhanced substantially by a higher accuracy in the numerical implementation of the aperture and rim functions. Additionally, the model's predictive power would be improved considerably by determining the wavelength- and probe-dependent complex factor C from measurements, physical principles or numerical simulations, instead of determining it on a phenomenological basis. Also, the influence of interactions between probe and sample on the calculated images could be studied in greater detail, as well as the possibility of resonances in the transmission characteristics of the probe. We believe that by identifying the corresponding coupling distributions, the imaging process of different types of aperture probes, such as the split-ring probe recently used by Burresi *et al.* [93], can be described by a similar model. In this way, near-field probes could also be specifically tailored to realise novel transfer functions.

Importantly, we can now compare experimentally determined aperture SNOM images of metamaterials to associated numerical calculations and learn more about the physical origin of the metamaterials' properties. Thus, we have paved the way for further detailed studies on a broad range of metamaterial structures. For this, nearly all types of planar metamaterial designs would be eligible. Also, more experiments on the spectral evolution of the eigenmodes in low-symmetry split-ring-resonator arrays and their excitation with polarisation directions other than the eigenpolarisations would certainly be interesting. In addition to this, a multitude of other nanostructured surfaces and near-field effects could be examined and compared to theory, particularly individual meta-atoms or small groups thereof, as well as interactions between them.

Bibliography

[1] V. G. Veselago, "The electrodynamics of substances with simultaneously negative values of ε and μ," Sov. Phys. Usp. **10**, 509–514 (1968).

[2] J. Pendry, A. Holden, D. Robbins, and W. Stewart, "Magnetism from conductors and enhanced nonlinear phenomena," IEEE Trans. Microwave Theory Tech. **47**, 2075–2084 (1999).

[3] M. Decker, S. Linden, and M. Wegener, "Coupling effects in low-symmetry planar split-ring resonator arrays," Opt. Lett. **34**, 1579–1581 (2009).

[4] S. Linden, M. Decker, and M. Wegener, "Model System for a One-Dimensional Magnetic Photonic Crystal," Phys. Rev. Lett. **97**, 083902 (2006).

[5] S. Zhang, W. Fan, K. J. Malloy, S. Brueck, N. C. Panoiu, and R. M. Osgood, "Near-infrared double negative metamaterials," Opt. Express **13**, 4922–4930 (2005).

[6] H. J. Lezec, J. A. Dionne, and H. A. Atwater, "Negative Refraction at Visible Frequencies," Science **316**, 430 (2007).

[7] J. B. Pendry, "Negative Refraction Makes a Perfect Lens," Phys. Rev. Lett. **85**, 3966–3969 (2000).

[8] S. Xi, H. Chen, T. Jiang, L. Ran, J. Huangfu, B.-I. Wu, J. A. Kong, and M. Chen, "Experimental Verification of Reversed Cherenkov Radiation in Left-Handed Metamaterial," Phys. Rev. Lett. **103**, 194801 (2009).

[9] J. B. Pendry, D. Schurig, and D. R. Smith, "Controlling Electromagnetic Fields," Science **312**, 1780–1782 (2006).

[10] U. Leonhardt, "Optical Conformal Mapping," Science **312**, 1777–1780 (2006).

[11] E. H. Synge, "A suggested method for extending microscopic resolution into the ultramicroscopic region," Philos. Mag. **6**, 356–362 (1928).

[12] H. A. Bethe, "Theory of Diffraction by Small Holes," Phys. Rev. **66**, 163–182 (1944).

[13] C. J. Bouwkamp, "On Bethe's theory of diffraction by small holes," Philips Res. Rep. **5**, 321–332 (1950).

[14] E. A. Ash and G. Nicholls, "Super-resolution Aperture Scanning Microscope," Nature **237**, 510–512 (1972).

[15] D. W. Pohl, W. Denk, and M. Lanz, "Optical stethoscopy: Image recording with resolution lambda/20," Appl. Phys. Lett. **44**, 651–653 (1984).

[16] E. Betzig, J. K. Trautman, T. D. Harris, J. S. Weiner, and R. L. Kostelak, "Breaking the Diffraction Barrier: Optical Microscopy on a Nanometric Scale," Science **251**, 1468–1470 (1991).

[17] G. A. Valaskovic, M. Holton, and G. H. Morrison, "Parameter control, characterization, and optimization in the fabrication of optical fiber near-field probes," Appl. Opt. **34**, 1215–1228 (1995).

[18] S. Mononobe and M. Ohtsu, "Fabrication of a pencil-shaped fiber probe for near-field optics by selective chemical etching," J. Lightwave Technol. **14**, 2231–2235 (1996).

[19] K. Karrai and R. D. Grober, "Piezoelectric tip-sample distance control for near field optical microscopes," Appl. Phys. Lett. **66**, 1842–1844 (1995).

[20] J. Wessel, "Surface-enhanced optical microscopy," J. Opt. Soc. Am. B **2**, 1538–1541 (1985).

[21] T. Zentgraf, J. Dorfmüller, C. Rockstuhl, C. Etrich, R. Vogelgesang, K. Kern, T. Pertsch, F. Lederer, and H. Giessen, "Amplitude- and phase-resolved optical near fields of split-ring-resonator-based metamaterials," Opt. Lett. **33**, 848–850 (2008).

[22] A. Bitzer, H. Merbold, A. Thoman, T. Feurer, H. Helm, and M. Walther, "Terahertz near-field imaging of electric and magnetic resonances of a planar metamaterial," Opt. Express **17**, 3826–3834 (2009).

[23] L. Novotny, D. W. Pohl, and P. Regli, "Light propagation through nanometer-sized structures: the two-dimensional-aperture scanning near-field optical microscope," J. Opt. Soc. Am. A **11**, 1768–1779 (1994).

[24] M. Totzeck, "Near-field imaging with a waveguide resonator probe in collection and reflection mode: II. Numerical simulation," J. Eur. Opt. Soc. Part A **3**, 879 (1994).

[25] J.-J. Greffet and R. Carminati, "Image formation in near-field optics," Prog. Surf. Sci. **56**, 133 – 237 (1997).

[26] S. I. Bozhevolnyi, B. Vohnsen, and E. A. Bozhevolnaya, "Transfer functions in collection scanning near-field optical microscopy," Opt. Commun. **172**, 171 – 179 (1999).

[27] E. Hecht, *Optik* (Oldenbourg Verlag, 2005).

[28] L. Novotny and B. Hecht, *Principles of Nano-Optics* (Cambridge University Press, 2008).

[29] I. Lindell, A. Sihvola, S. Tretyakov, and A. Viitanen, *Electromagnetic Waves in Chiral and Bi-Isotropic Media* (Artech House, 1994).

[30] P. K. L. Drude, "Zur Elektronentheorie der Metalle," Ann. Phys. **306**, 566–613 (1990).

[31] P. B. Johnson and R. W. Christy, "Optical Constants of the Noble Metals," Phys. Rev. B **6**, 4370–4379 (1972).

[32] S. Linden, C. Enkrich, M. Wegener, J. Zhou, T. Koschny, and C. M. Soukoulis, "Magnetic Response of Metamaterials at 100 Terahertz," Science **306**, 1351–1353 (2004).

[33] A. N. Grigorenko, A. K. Geim, H. F. Gleeson, Y. Zhang, A. A. Firsov, I. Y. Khrushchev, and J. Petrovic, "Nanofabricated media with negative permeability at visible frequencies," Nature **438**, 335–338 (2005).

[34] D. R. Smith, W. J. Padilla, D. C. Vier, S. C. Nemat-Nasser, and S. Schultz, "Composite Medium with Simultaneously Negative Permeability and Permittivity," Phys. Rev. Lett. **84**, 4184–4187 (2000).

[35] R. A. Shelby, D. R. Smith, and S. Schultz, "Experimental Verification of a Negative Index of Refraction," Science **292**, 77–79 (2001).

[36] D. R. Smith, J. B. Pendry, and M. C. K. Wiltshire, "Metamaterials and Negative Refractive Index," Science **305**, 788–792 (2004).

[37] S. Linden, C. Enkrich, G. Dolling, M. W. Klein, J. Zhou, T. Koschny, C. M. Soukoulis, S. Burger, F. Schmidt, and M. Wegener, "Photonic Metamaterials: Magnetism at Optical Frequencies," IEEE J. Sel. Top. Quantum Electron. **12**, 1097–1105 (2006).

[38] S. Zhang, W. Fan, N. C. Panoiu, K. J. Malloy, R. M. Osgood, and S. R. J. Brueck, "Experimental Demonstration of Near-Infrared Negative-Index Metamaterials," Phys. Rev. Lett. **95**, 137404 (2005).

[39] W. N. Hardy and L. A. Whitehead, "Split-ring resonator for use in magnetic resonance from 200–2000 MHz," Rev. Sci. Instrum. **52**, 213–216 (1981).

[40] N.-A. Feth, "Nonlinear Optics of Planar Metamaterial Arrays and Spectroscopy of Individual "Photonic Atoms"," Ph.D. thesis, Universität Karlsruhe (TH), Institut für Angewandte Physik (2010).

[41] M. Decker, "New Light on Optical Activity: Interaction of Electromagnetic Waves with Chiral Photonic Metamaterials," Ph.D. thesis, Universität Karlsruhe (TH), Institut für Angewandte Physik (2010).

[42] M. W. Klein, C. Enkrich, M. Wegener, C. M. Soukoulis, and S. Linden, "Single-slit split-ring resonators at optical frequencies: limits of size scaling," Opt. Lett. **31**, 1259–1261 (2006).

[43] J. Zhou, T. Koschny, M. Kafesaki, E. N. Economou, J. B. Pendry, and C. M. Soukoulis, "Saturation of the Magnetic Response of Split-Ring Resonators at Optical Frequencies," Phys. Rev. Lett. **95**, 223902 (2005).

[44] R. W. Wood, "On a remarkable case of uneven distribution of light in a diffraction grating spectrum," Philos. Mag. **4**, 396 (1902).

[45] J. W. Strutt (Lord Rayleigh), "On the Dynamical Theory of Gratings," Proc. R. Soc. London, Ser. A **79**, 399–416 (1907).

[46] S. Linden, N. Rau, U. Neuberth, A. Naber, M. Wegener, S. Pereira, K. Busch, A. Christ, and J. Kuhl, "Near-field optical microscopy and spectroscopy of one-dimensional metallic photonic crystal slabs," Phys. Rev. B **71**, 245119 (2005).

[47] G. Dolling, C. Enkrich, M. Wegener, J. F. Zhou, C. M. Soukoulis, and S. Linden, "Cut-wire pairs and plate pairs as magnetic atoms for optical metamaterials," Opt. Lett. **30**, 3198–3200 (2005).

[48] G. Dolling, M. Wegener, C. M. Soukoulis, and S. Linden, "Negative-index metamaterial at 780 nm wavelength," Opt. Lett. **32**, 53–55 (2007).

[49] G. Dolling, "Design, Fabrication, and Characterization of Double-Negative Metamaterials for Photonics," Ph.D. thesis, Universität Karlsruhe (TH), Institut für Angewandte Physik (2007).

[50] G. Dolling, M. Wegener, A. Schadle, S. Burger, and S. Linden, "Observation of magnetization waves in negative-index photonic metamaterials," Appl. Phys. Lett. **89**, 231118 (2006).

[51] N. Liu, H. Guo, L. Fu, S. Kaiser, H. Schweizer, and H. Giessen, "Three-dimensional photonic metamaterials at optical frequencies," Nat. Mater. **7**, 31 (2007).

[52] N. Liu and H. Giessen, "Three-dimensional optical metamaterials as model systems for longitudinal and transverse magnetic coupling," Opt. Express **16**, 21233–21238 (2008).

[53] N. Liu, S. Kaiser, and H. Giessen, "Magnetoinductive and Electroinductive Coupling in Plasmonic Metamaterial Molecules," Adv. Mater. **20**, 4521–4525 (2008).

[54] N. Liu, H. Liu, S. Zhu, and H. Giessen, "Stereometamaterials," Nat. Photonics **3**, 157–162 (2009).

[55] K. Yee, "Numerical solution of initial boundary value problems involving maxwell's equations in isotropic media," IEEE Trans. Antennas Propag. **14**, 302–307 (1966).

[56] T. Roesener, "Nahfeldmikroskopie an 3D Photonischen Kristallen," Diplom Thesis, Universität Karlsruhe (TH), Institut für Angewandte Physik (2007).

[57] J. W. Strutt (Lord Rayleigh), "Investigations in optics with special reference to the spectroscope." Philos. Mag. **5**, 261 (1879).

[58] E. Abbe, "Beiträge zur Theorie des Mikroskops und der mikroskopischen Wahrnehmung." Arch. mikroskop. Anat. **9**, 413 (1873).

[59] S. W. Hell, "Far-Field Optical Nanoscopy," Science **316**, 1153–1158 (2007).

[60] J. M. Vigoureux, F. Depasse, and C. Girard, "Superresolution of near-field optical microscopy defined from properties of confined electromagnetic waves," Appl. Opt. **31**, 3036–3045 (1992).

[61] J. M. Vigoureux and D. Courjon, "Detection of nonradiative fields in light of the Heisenberg uncertainty principle and the Rayleigh criterion," Appl. Opt. **31**, 3170–3177 (1992).

[62] B. Hecht, B. Sick, U. P. Wild, V. Deckert, R. Zenobi, O. J. F. Martin, and D. W. Pohl, "Scanning near-field optical microscopy with aperture probes: Fundamentals and applications," J. Chem. Phys. **112**, 7761–7774 (2000).

[63] R. Wiesendanger, ed., *Scanning Probe Microscopy - Analytical Methods* (Springer, 1998).

[64] G. v. Freymann, "Simulationen zur optischen Rasternahfeldmikroskopie," Diplom Thesis, Universität Karlsruhe (TH), Institut für Angewandte Physik (1997).

[65] T. Taubner, "Optische Rasternahfeldmikroskopie mittels Tapping-SNOM," Diplom Thesis, Universität Karlsruhe, Institut für Angewandte Physik (2001).

[66] R. Hillenbrand, B. Knoll, and F. Keilmann, "Pure optical contrast in scattering-type scanning near-field microscopy," J. Microsc. **202**, 77–83 (2001).

[67] J. Koglin, U. C. Fischer, and H. Fuchs, "Material contrast in scanning near-field optical microscopy at 1–10 nm resolution," Phys. Rev. B **55**, 7977–7984 (1997).

[68] B. Knoll, F. Keilmann, A. Kramer, and R. Guckenberger, "Contrast of microwave near-field microscopy," Appl. Phys. Lett. **70**, 2667–2669 (1997).

[69] R. Stöckle, C. Fokas, V. Deckert, R. Zenobi, B. Sick, B. Hecht, and U. P. Wild, "High-quality near-field optical probes by tube etching," Appl. Phys. Lett. **75**, 160–162 (1999).

[70] A. Naber, D. Molenda, U. C. Fischer, H.-J. Maas, C. Höppener, N. Lu, and H. Fuchs, "Enhanced light confinement in a near-field optical probe with a triangular aperture," Phys. Rev. Lett. **89**, 210801 (2002).

[71] D. Molenda, G. C. des Francs, U. C. Fischer, N. Rau, and A. Naber, "High-resolution mapping of the optical near-field components at a triangular nano-aperture," Opt. Express **13**, 10688–10696 (2005).

[72] F. de Lange, A. Cambi, R. Huijbens, B. de Bakker, W. Rensen, M. Garcia-Parajo, N. van Hulst, and C. G. Figdor, "Cell biology beyond the diffraction limit: near-field scanning optical microscopy," J. Cell Sci. **114**, 4153–4160 (2001).

[73] B. Hecht, H. Bielefeldt, Y. Inouye, D. W. Pohl, and L. Novotny, "Facts and artifacts in near-field optical microscopy," J. Appl. Phys. **81**, 2492–2498 (1997).

[74] H. C. Adelmann, "Experimente zur optischen Depolarisations–Rasternahfeldmikroskopie," Diplom Thesis, Universität Karlsruhe (TH), Institut für Angewandte Physik (1998).

[75] M. Muranishi, K. Sato, S. Hosaka, A. Kikukawa, T. Shintani, and K. Ito, "Control of Aperture Size of Optical Probes for Scanning Near-Field Optical Microscopy Using Focused Ion Beam Technology," Jpn. J. Appl. Phys., Part 2 **36**, L942–L944 (1997).

[76] A. A. Tseng, "Recent developments in micromilling using focused ion beam technology," J. Micromech. Microeng. **14**, R15 (2004).

[77] G. v. Freymann, "Der Einfluss von Unordnung auf die Energieniveaustatistik von Exzitonen in Halbleitern." Ph.D. thesis, Universität Karlsruhe (TH), Institut für Angewandte Physik (2001).

[78] U. Neuberth, "Optische Rasternahfeldmikroskopie und -spektroskopie an Halbleiternanostrukturen und Plasmonischen Kristallen," Ph.D. thesis, Universität Karlsruhe (TH), Institut für Angewandte Physik (2004).

[79] M. Burresi, D. Diessel, D. van Oosten, S. Linden, M. Wegener, and L. Kuipers, "Negative-Index Metamaterials: Looking into the Unit Cell," Nano Lett. **10**, 2480–2483 (2010).

[80] D. Diessel, M. Decker, S. Linden, and M. Wegener, "Near-field optical experiments on low-symmetry split-ring-resonator arrays," Opt. Lett. **35**, 3661–3663 (2010).

[81] C. Enkrich, "Magnetic Metamaterials for Photonics," Ph.D. thesis, Universität Karlsruhe (TH), Institut für Angewandte Physik (2006).

[82] M. Deubel, "Three-Dimensional Photonic Crystals via Direct Laser Writing: Fabrication and Characterization," Ph.D. thesis, Universität Karlsruhe (TH), Institut für Angewandte Physik (2006).

[83] M. Sandtke, R. J. P. Engelen, H. Schoenmaker, I. Attema, H. Dekker, I. Cerjak, J. P. Korterik, F. B. Segerink, and L. Kuipers, "Novel instrument for surface plasmon polariton tracking in space and time," Rev. Sci. Instrum. **79**, 013704 (2008).

[84] M. Burresi, "Nanoscale investigation of light-matter interactions mediated by magnetic and electric coupling," Ph.D. thesis, University of Twente (2009).

[85] C. Girard and A. Dereux, "Near-field optics theories," Rep. Prog. Phys. **59**, 657–699 (1996).

[86] A. Castiaux, C. Girard, M. Spajer, and S. Davy, "Near-field optical effects inside a photosensitive sample coupled with a SNOM tip," Ultramicroscopy **71**, 49 – 58 (1998).

[87] P. J. Valle, J.-J. Greffet, and R. Carminati, "Optical contrast, topographic contrast and artifacts in illumination-mode scanning near-field optical microscopy," J. Appl. Phys. **86**, 648–656 (1999).

[88] M. Tanaka and K. Tanaka, "Computer simulation for two-dimensional near-field optics with use of a metal-coated dielectric probe," J. Opt. Soc. Am. A **18**, 919–925 (2001).

[89] M. A. Ordal, L. L. Long, R. J. Bell, S. E. Bell, R. R. Bell, J. R. W. Alexander, and C. A. Ward, "Optical properties of the metals Al, Co, Cu, Au, Fe, Pb, Ni, Pd, Pt, Ag, Ti, and W in the infrared and far infrared," Appl. Opt. **22**, 1099–1119 (1983).

[90] N. A. Rau, "Experimente zur optischen Nahfeldmikroskopie an eindimensionalen metallischen Photonischen Kristallen," Diplom Thesis, Universität Karlsruhe (TH), Institut für Angewandte Physik (2003).

[91] G. Dolling, C. Enkrich, M. Wegener, C. M. Soukoulis, and S. Linden, "Simultaneous Negative Phase and Group Velocity of Light in a Metamaterial," Science **312**, 892–894 (2006).

[92] G. Dolling, C. Enkrich, M. Wegener, C. M. Soukoulis, and S. Linden, "Low-loss negative-index metamaterial at telecommunication wavelengths," Opt. Lett. **31**, 1800–1802 (2006).

[93] M. Burresi, D. van Oosten, T. Kampfrath, H. Schoenmaker, R. Heideman, A. Leinse, and L. Kuipers, "Probing the Magnetic Field of Light at Optical Frequencies," Science **326**, 550–553 (2009).

Publications

Parts of this thesis have already been published in scientific journals:

- M. Burresi, D. Diessel, D. van Oosten, S. Linden, M. Wegener, and L. Kuipers, "Negative-Index Metamaterials: Looking into the Unit Cell," Nano Lett. **10**, 2480–2483 (2010).
- D. Diessel, M. Decker, S. Linden, and M. Wegener, "Near-field optical experiments on low-symmetry split-ring-resonator arrays," Opt. Lett. **35**, 3661–3663 (2010).

Acknowledgements

At this point I would like to thank those who, with their assistance, contributed to this thesis.

First of all I thank my supervisor Prof. Dr. Martin Wegener for the opportunity to conduct such interesting research. Without the use of his group's excellent equipment, without the many enlightening discussions and meetings, and without his invaluable contributions to our publications the successful completion of this work would hardly have been possible. Secondly, I thank Prof. Dr. Kurt Busch, who kindly agreed to referee this thesis. I am also highly grateful to Prof. Dr. Stefan Linden for the funding of my Ph.D. within the Helmholtz-Hochschul-Nachwuchsgruppe programme, for innumerable helpful suggestions and advice, for the fabrication of the double-wire samples, and for his help with both the technical aspects of the simulations and the physical details of the probe model. And I acknowledge the DFG Center for Functional Nanostructures and the Karlsruhe School of Optics and Photonics for financial support.

To Prof. Dr. Laurens (Kobus) Kuipers, Dr. Matteo Burresi and Dr. Dries van Oosten I am indebted for some decisive insights into the workings of near-field probes. I particularly, and warmly, recognise Matteo Burresi, who carried out the phase- and polarisation-sensitive near-field measurements. Thanks also to Hinco Schoenmaker for fabricating both the usual near-field probes and the protrusion probes.

Furthermore, I gratefully acknowledge the always good-natured helpfulness of Prof. Dr. Georg von Freymann, who assisted me many times with the running of the near-field microscope, and with handling the hydrofluoric acid. I thank Dr. Manuel Decker for the design and fabrication of the low-symmetry split-ring-resonator array samples, as well as for many valuable discussions, and Dr. Gunnar Dolling for the use of his fishnet structures and the associated characterisation data.

My gratitude is also due to: Patrice Brenner, the focused-ion-beam technician, who executed the aperture cutting of the near-field probes; Jacques Hawecker for taking hundreds of SEM images of probes and samples; both the current and the former technical staff of our group and that of our institute's electronic workshop, who solved a myriad of minor and major problems and were always happy to help; and the always supportive secretaries of the Institut für Angewandte Physik. Additionally, I thank all my colleagues for the congenial spirit in our group. It was a real pleasure to work with you.

An important contribution was the diligent proofreading by Stefan Linden, Manuel Decker, Monika Diessel, Julia Budde, Veronika Haug and Georg von Freymann. Thank you all.

Finally, I thank my family, friends and colleagues for their support. It is greatly appreciated, thank you.

I want morebooks!

Buy your books fast and straightforward online - at one of world's fastest growing online book stores! Environmentally sound due to Print-on-Demand technologies.

Buy your books online at
www.morebooks.shop

Kaufen Sie Ihre Bücher schnell und unkompliziert online – auf einer der am schnellsten wachsenden Buchhandelsplattformen weltweit! Dank Print-On-Demand umwelt- und ressourcenschonend produziert.

Bücher schneller online kaufen
www.morebooks.shop

KS OmniScriptum Publishing
Brivibas gatve 197
LV-1039 Riga, Latvia
Telefax: +371 686 204 55

info@omniscriptum.com
www.omniscriptum.com

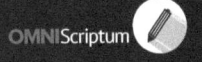

Printed by Books on Demand GmbH, Norderstedt / Germany